A MIDSHIPMAN'S JOURNAL

A reflection of those
four years by the bay

the story of

____

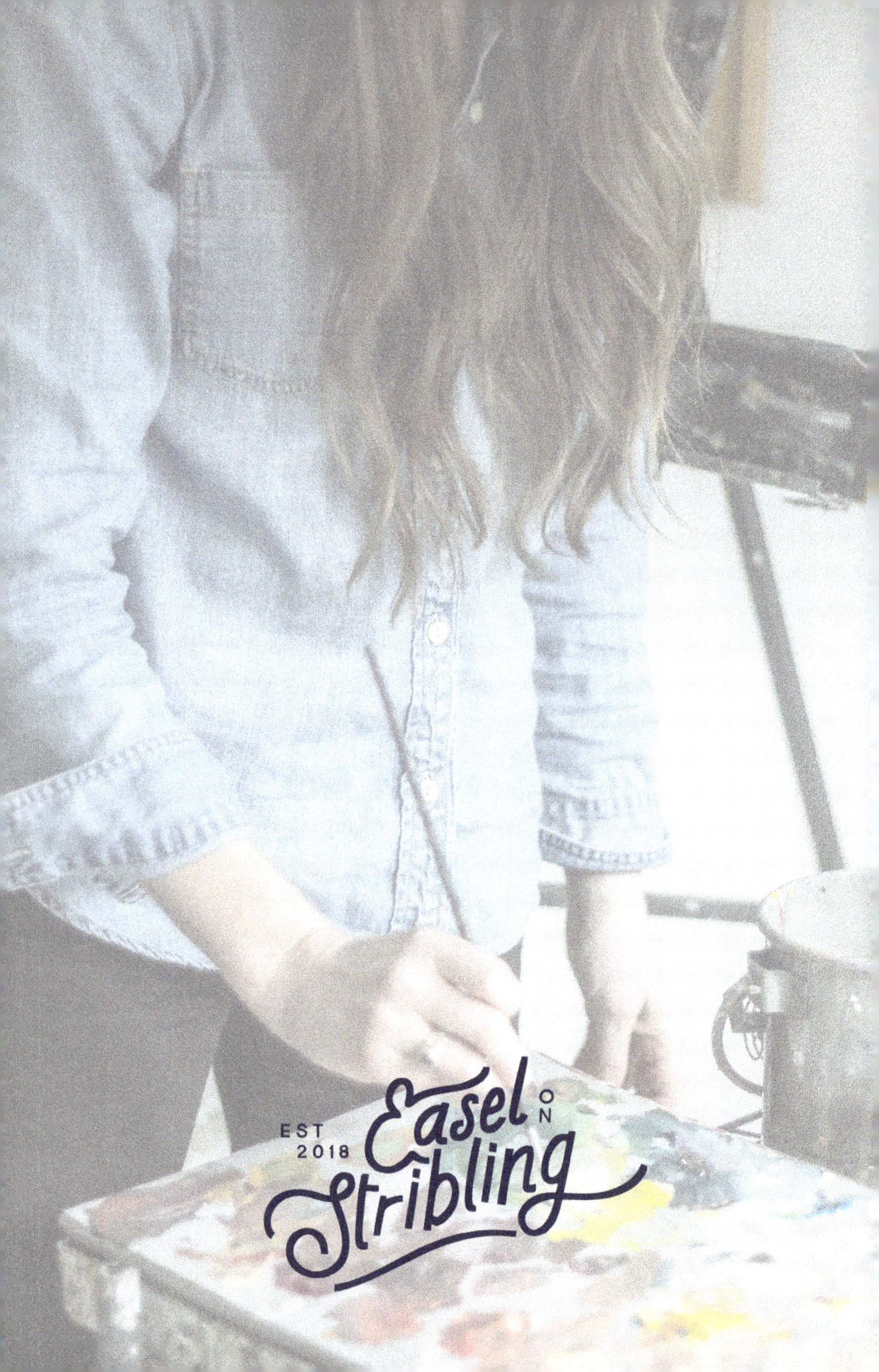

EST
2018
Easel
on
Stribling

ABOUT THIS BOOK

Since 2018, I have been working my way through a series of paintings exploring moments, big and small, of the United States Naval Academy. I use the pseudonym "Easel on Stribling," fondly named after the brick pathway each midshipman knows intimately. In my artwork, I am drawn to the common story, the one that each and every midshipman can related to.

As I painted, I tried to keep the moments vague and relatable. But, as I worked through them, I began to realize that inevitably, my own personal perspective emerged.

I resisted it at first: It was not about me, it is about you. Also, my story is not particularly extraordinary. I'm not an astronaut, I never wore more than a single stripe, I fell asleep in a lot of atypical places. I studied pretty hard and had just a couple close friends, but was mostly another face in the crowd and a frequent visitor to 7th wing gym. I was ordinary.

Then, it occured me, that story matters, too. And so does yours.

Little by little, I would paint a moment and remember my own memories of it. Some made me laugh, others made me cry. Some, in retrospect, brought up feelings of anger or embarrassment. Others, fond nostalgia. I wanted to find a way to invite you to reflect the way painting allowed me to.

This notebook was thus born. It was made in conjunction with a series of paintings exhibited at the Naval Academy Museum in 2021-2022 titled "A Midshipman's Journey." Each painting, including a few additional pieces not included in the exhibition, will be included throughout the journal to help guide you through the question prompts.

Beat Army!
Kristin (Otterson) Cronic, USNA Class of 2011
www.easelonstribling.com

How to use this book

Your story matters.

This notebook is a place to remember your story of the United States Naval Academy. If you have memories of Stribling Walk, chow calls, and King Hall...this is for you.

I hope you use this to remember the details of your experience: the moments that made you laugh, the ones that surprised you, and even the ones that may not have gone so great.

Pick it up and answer just one question every so often. or fill out the entire book in a day. There is no wrong way to capture your story.

You can share this book with future generations, use it to connect with old shipmates, or just to reminisce.

Enjoy!

Let's get started!

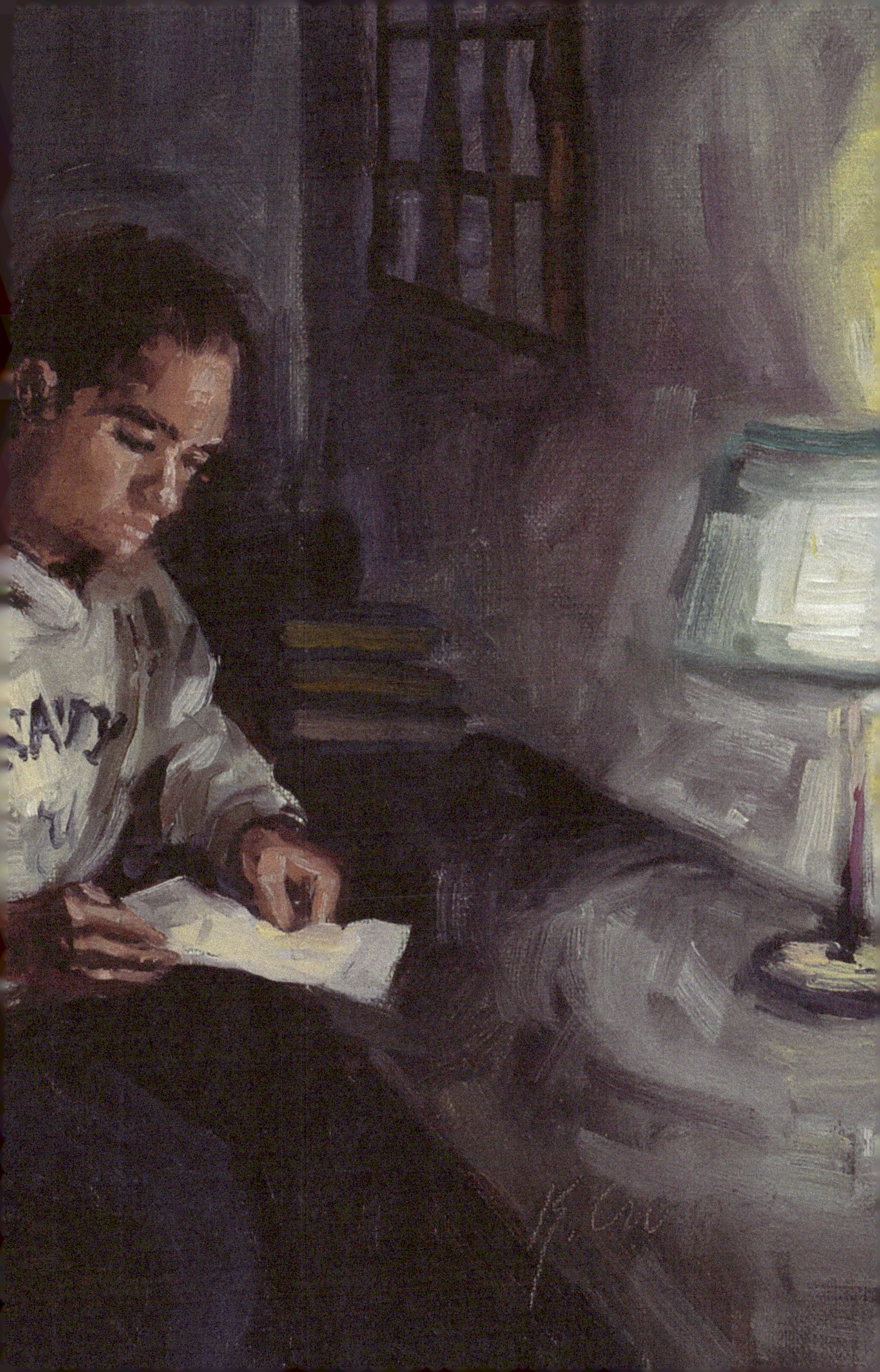

What were the circumstances around receiving your acceptance letter?

How did you feel when you received the envelope?

How did your loved ones respond?

What was your plan "B" ?

What made you want to apply to begin with?

Did anyone in your family serve in the military?

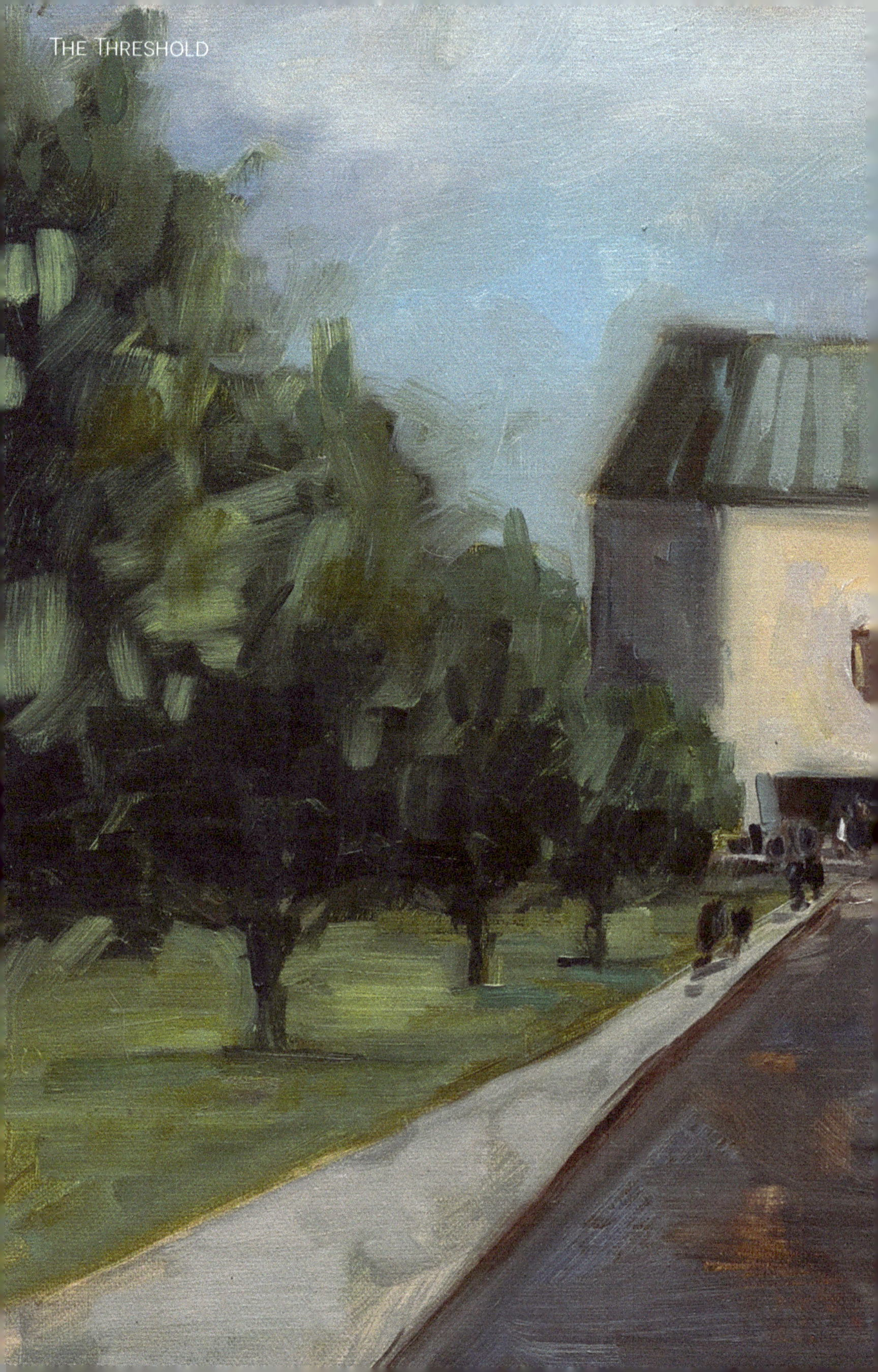
THE THRESHOLD

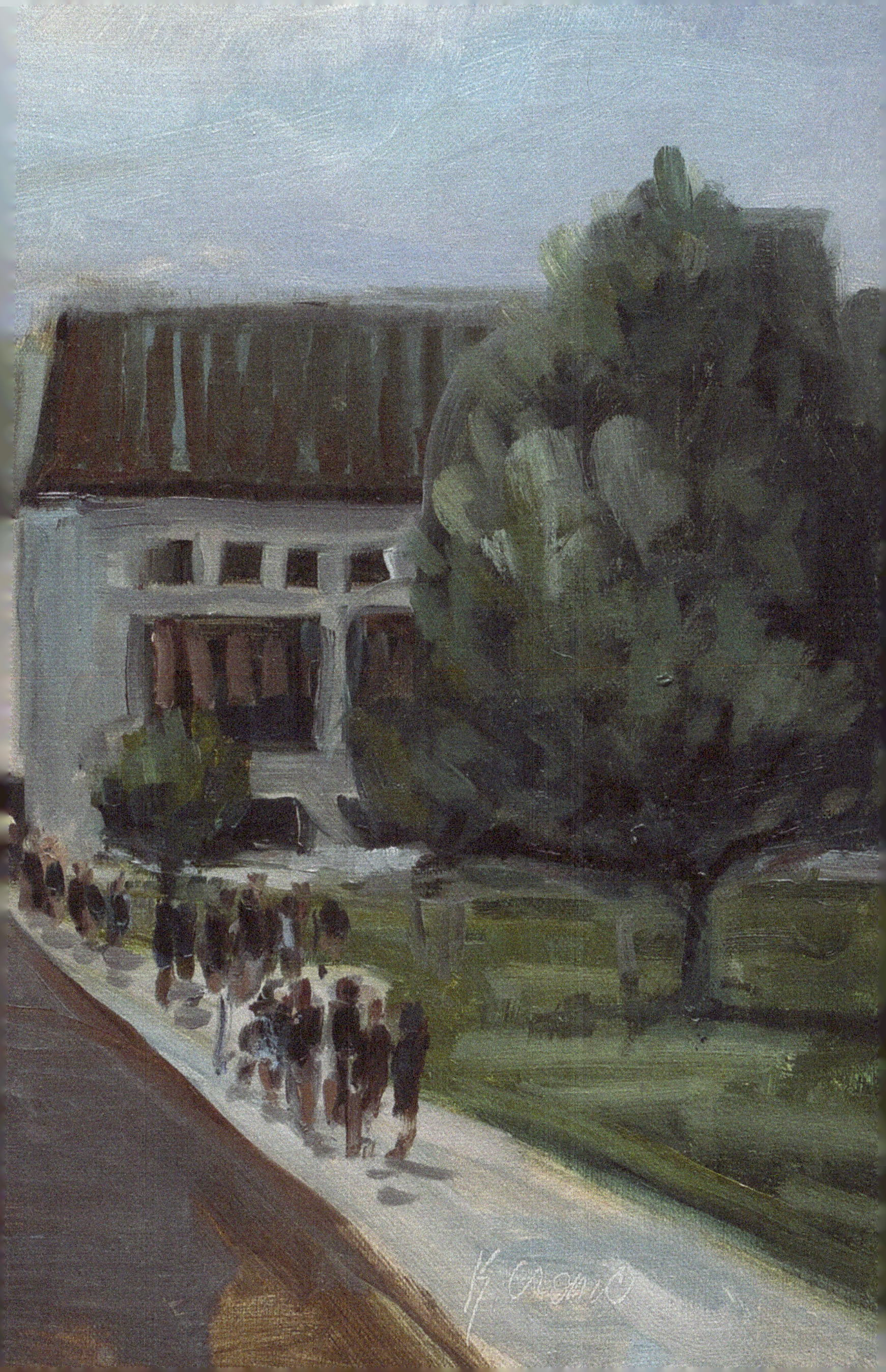

What are your strongest memories of I-day?

Who was with you? Where did you stay before hand?

Do you remember anyone you met during that day?

How did you prepare for plebe summer?

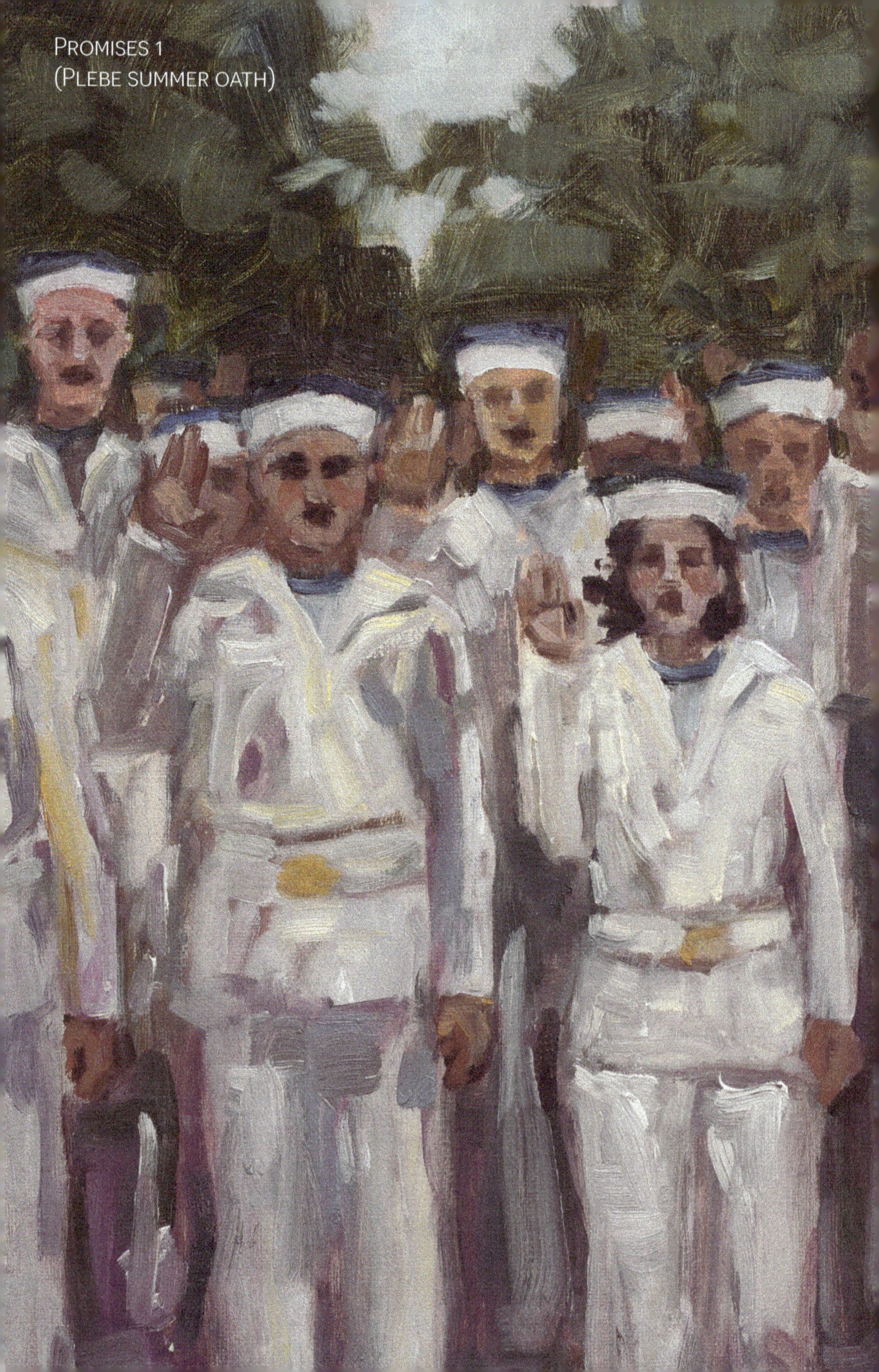
Promises 1
(Plebe summer oath)

Describe one funny memory from plebe summer.

What was your biggest frustration from plebe summer?

What was the worst punishment you had to serve and why?

Who did you room with during plebe summer? Who were your squad leaders and platoon commanders?

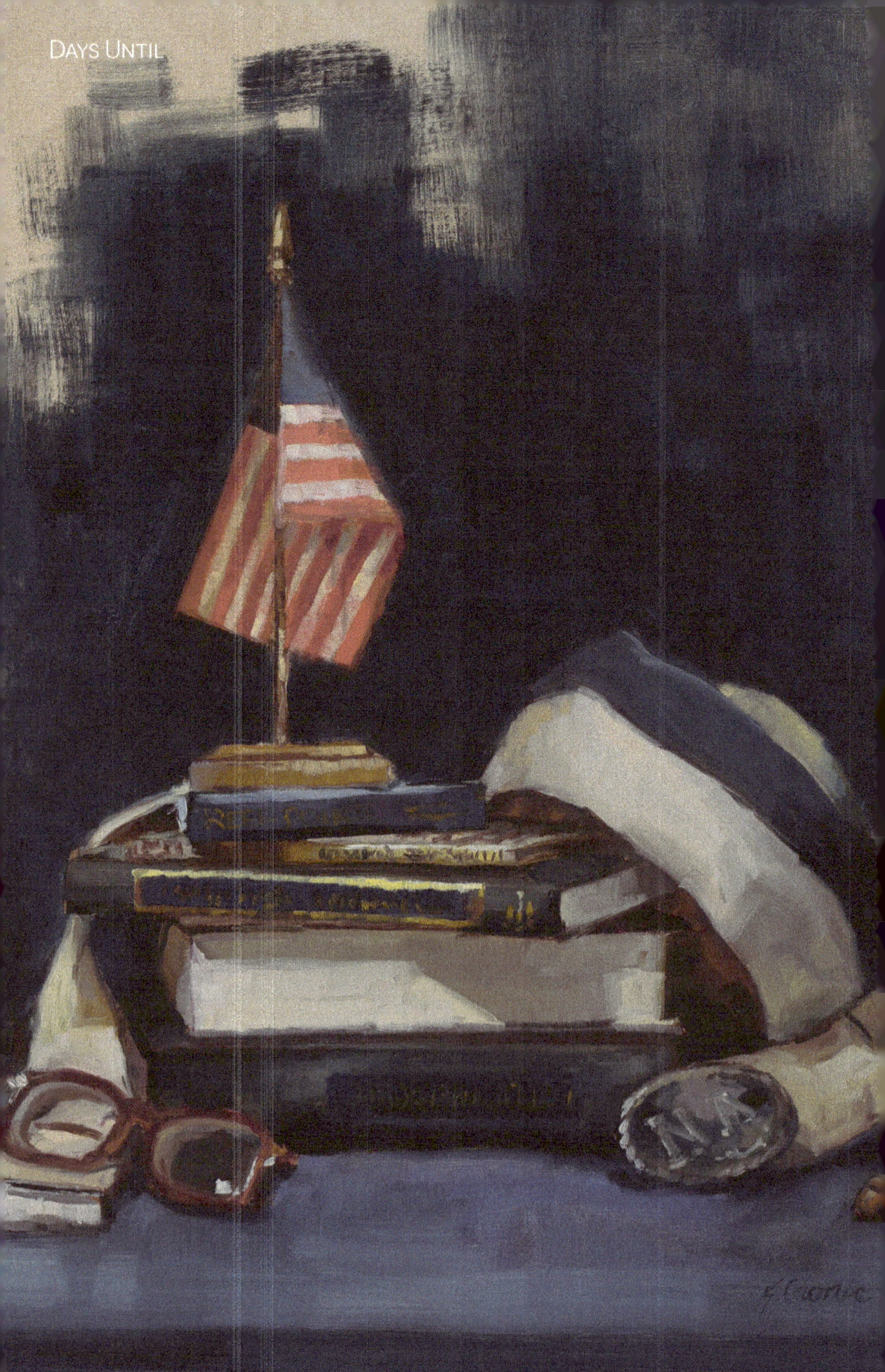
Days Until

What are your memories of Plebe Parents Weekend?

Who were your sponsor parents? (assigned or adopted)

Who did you room with plebe year?

Write down your memories from Hello Night.

Share a memorable bet you or a classmate wagered.

What new things did a company mate teach you?

What were your most challenging and your favorite classes
plebe year?

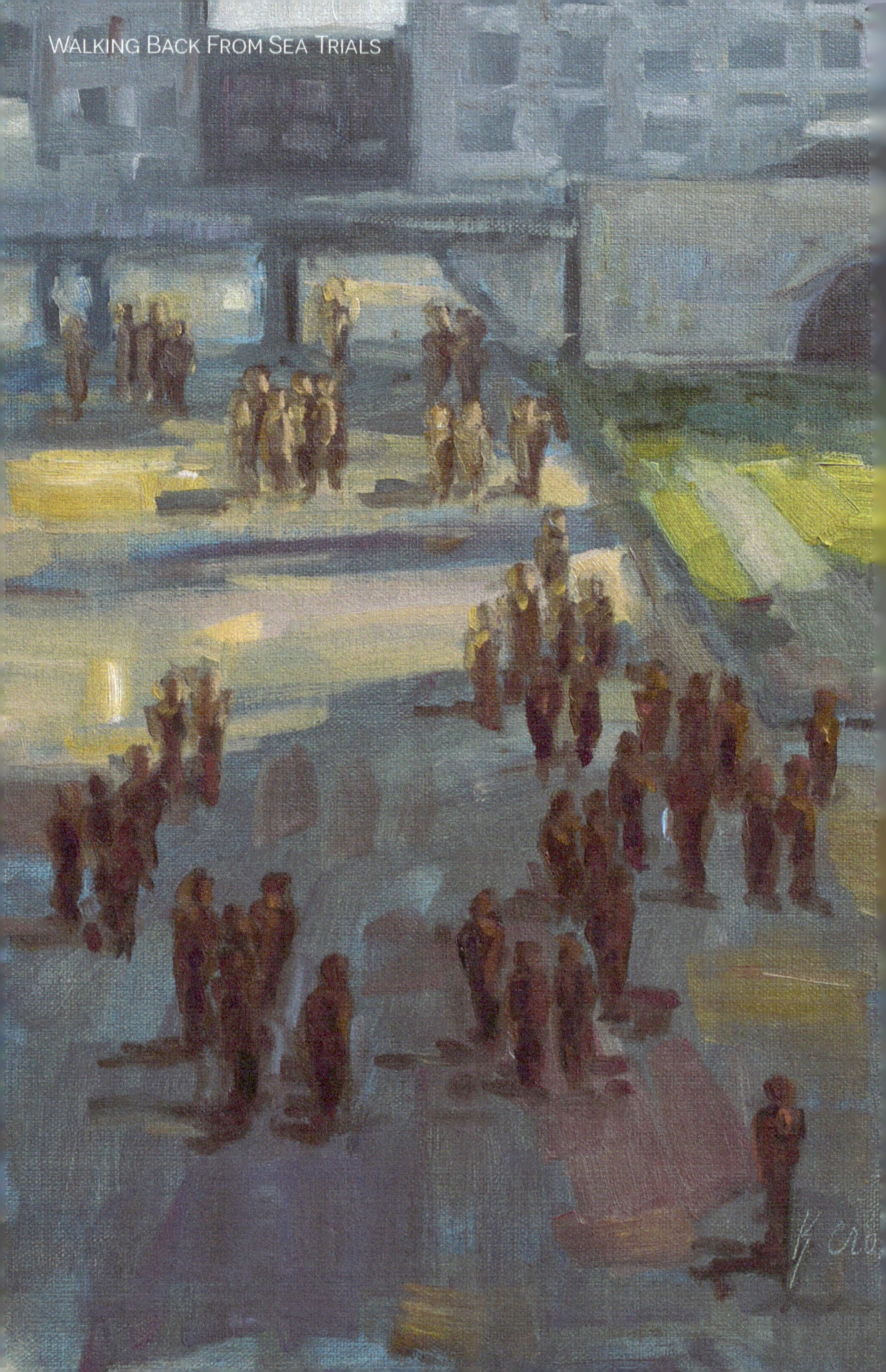
WALKING BACK FROM SEA TRIALS

Write down your memories of Sea Trials. What was the hardest
part of the day?

What was the best part of the day?

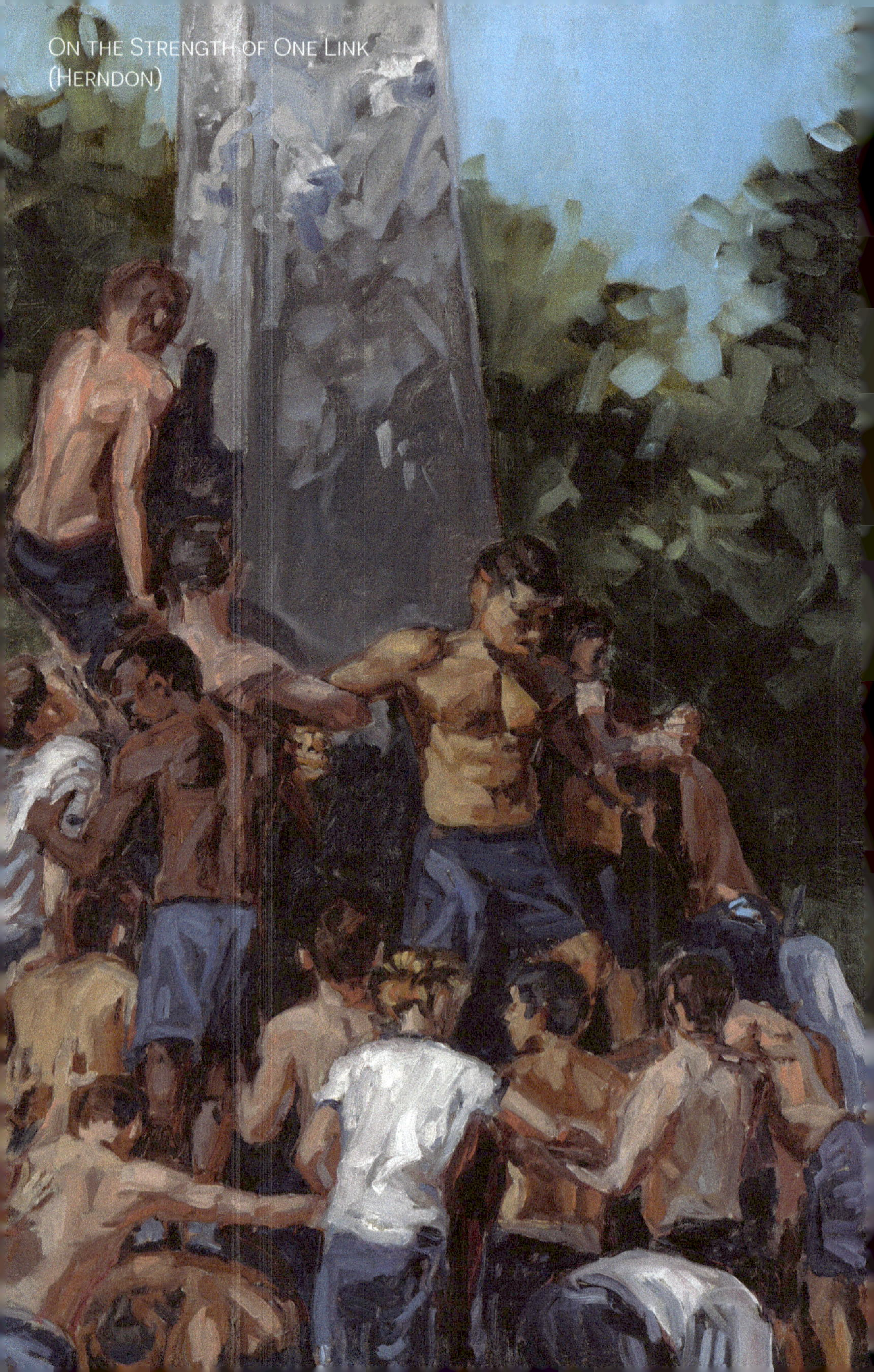
On the Strength of One Link
(Herndon)

What was Herndon like? Who were you with? What was the

atmosphere in company area like before and after? Did you have

fun? How did you celebrate?

How long did it take your class?

Use this space to write down any additional plebe memories that come to mind.

What did you do for summer training as a youngster? Share a memory from each training block (leave included!).

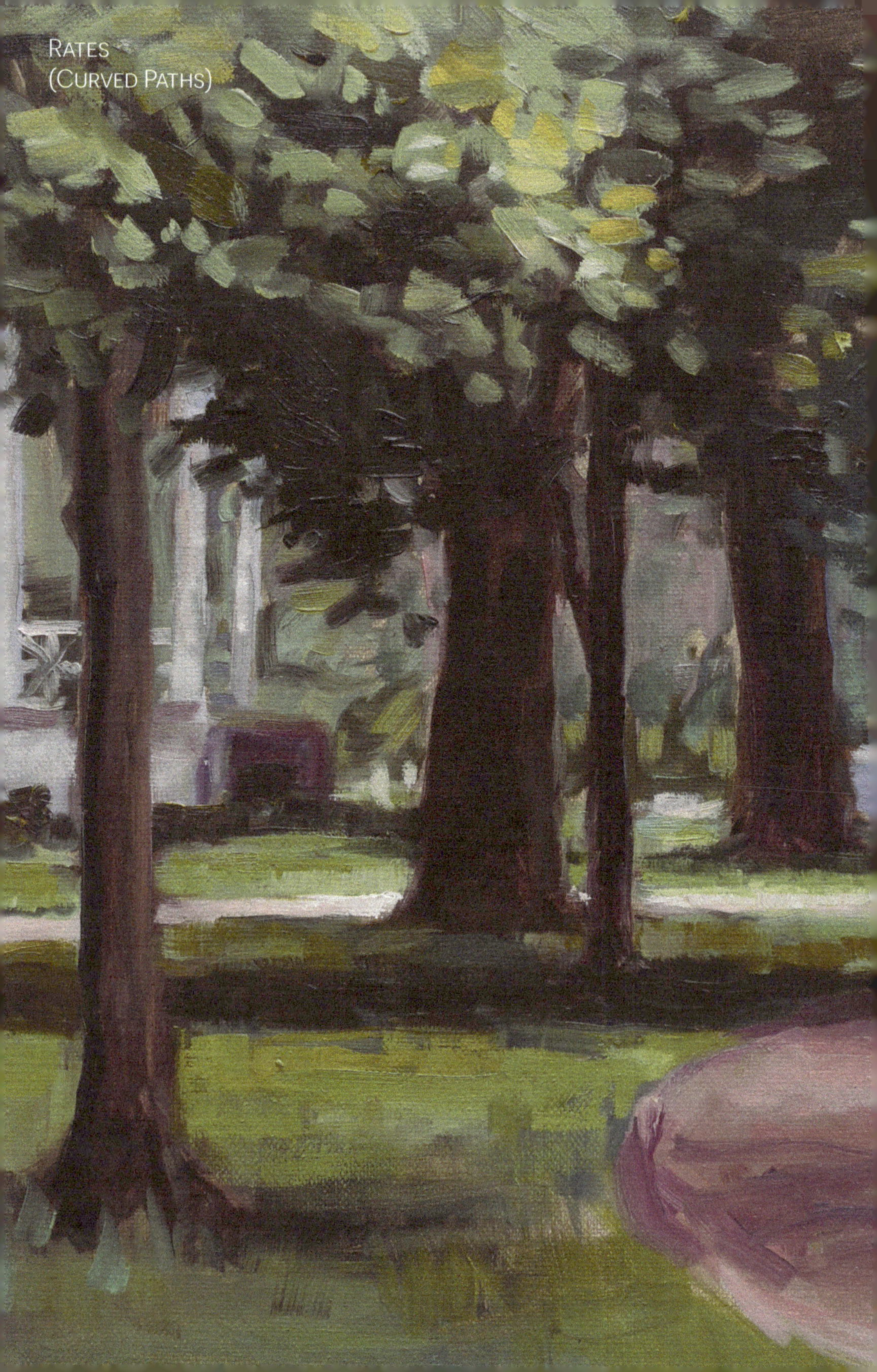
Rates
(Curved Paths)

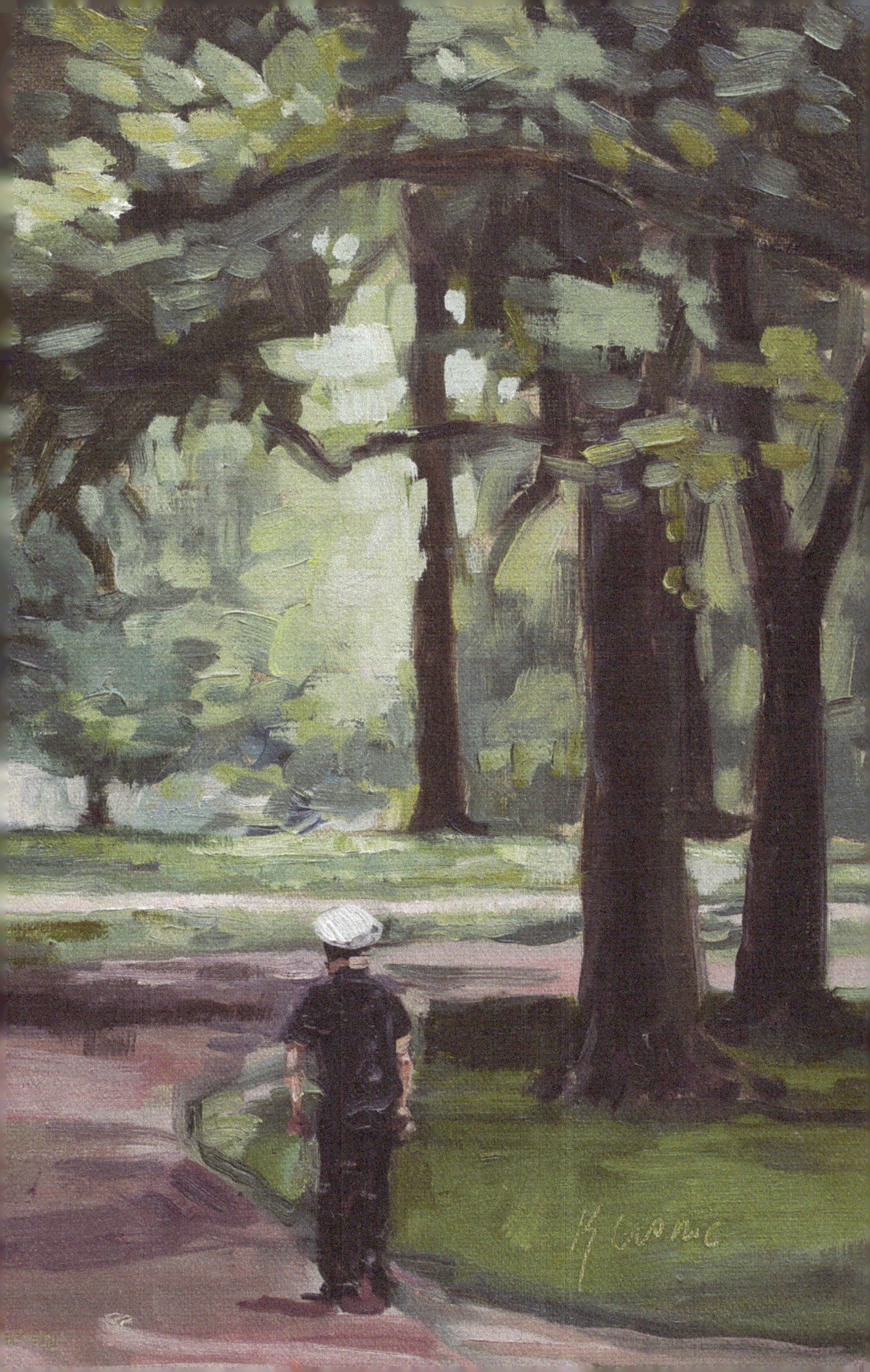

What freedom were you most excited about as a Youngster?

What ECA's were you involved with during your time at USNA?

What was a surprise or a great memory from Youngster year?

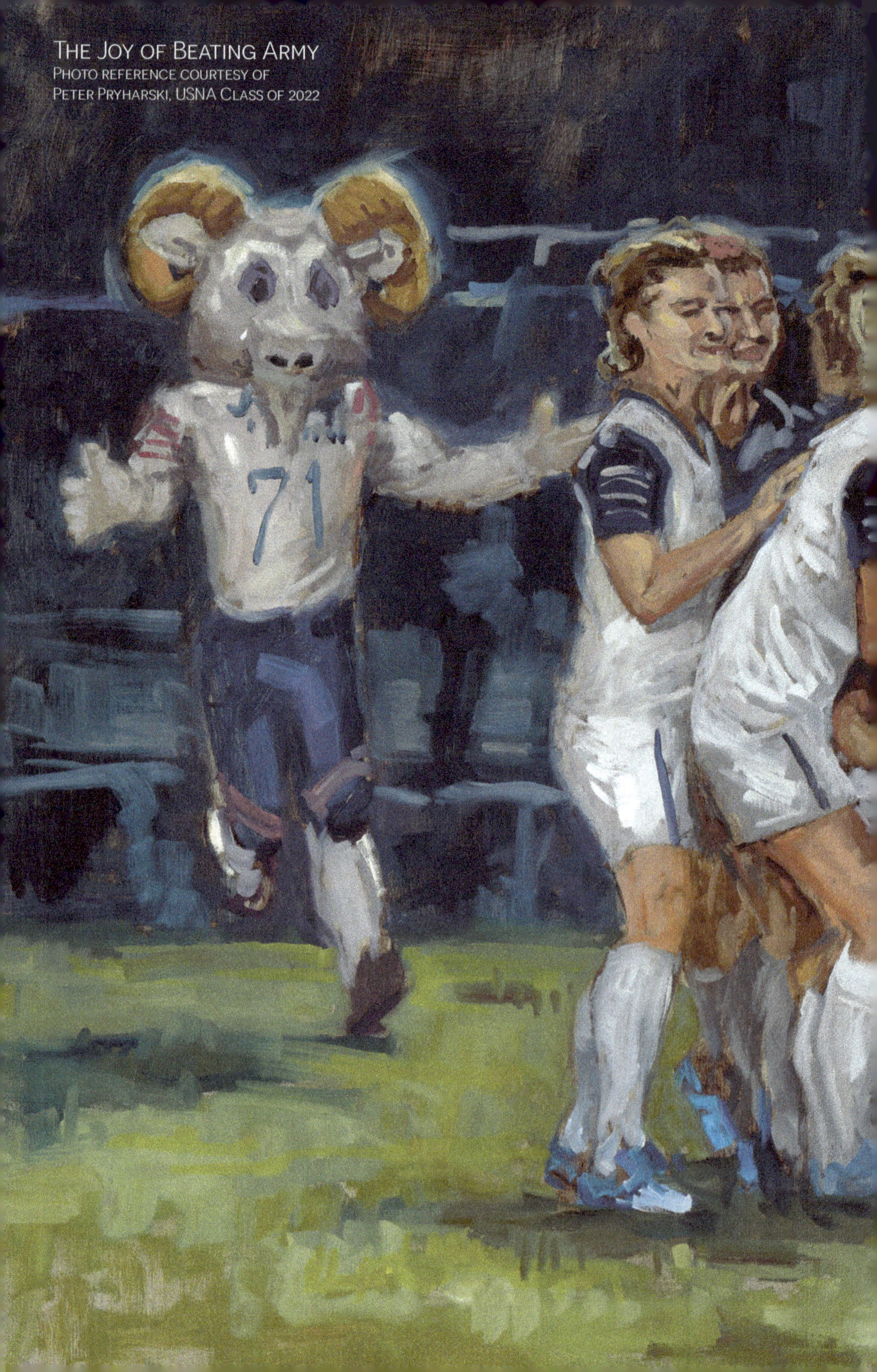
The Joy of Beating Army
Photo reference courtesy of
Peter Pryharski, USNA Class of 2022

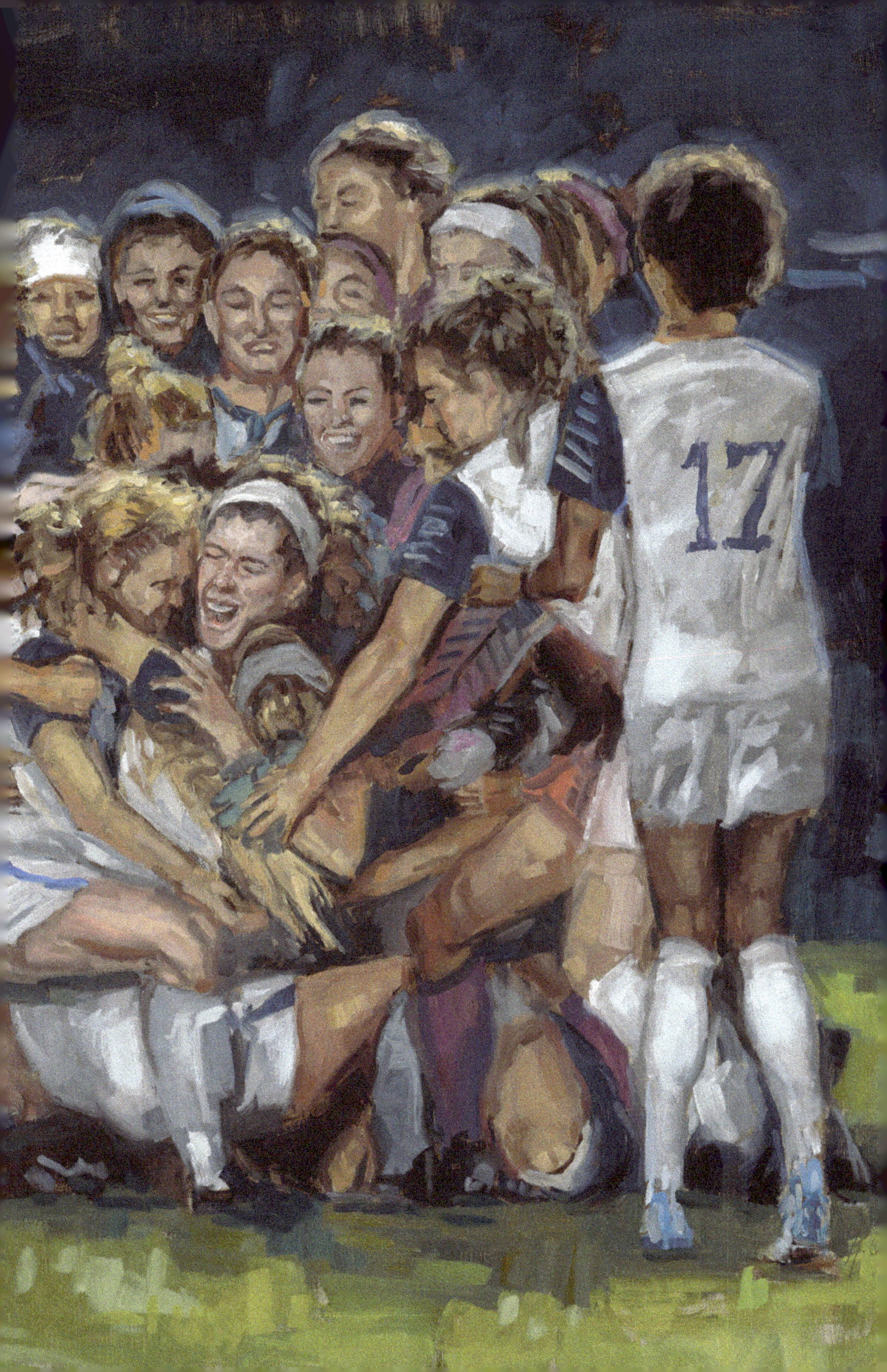

What sports were you involved with during your time at USNA?

Did you ever compete against Army? What was the outcome?

Share your favorite memory from athletics at USNA.

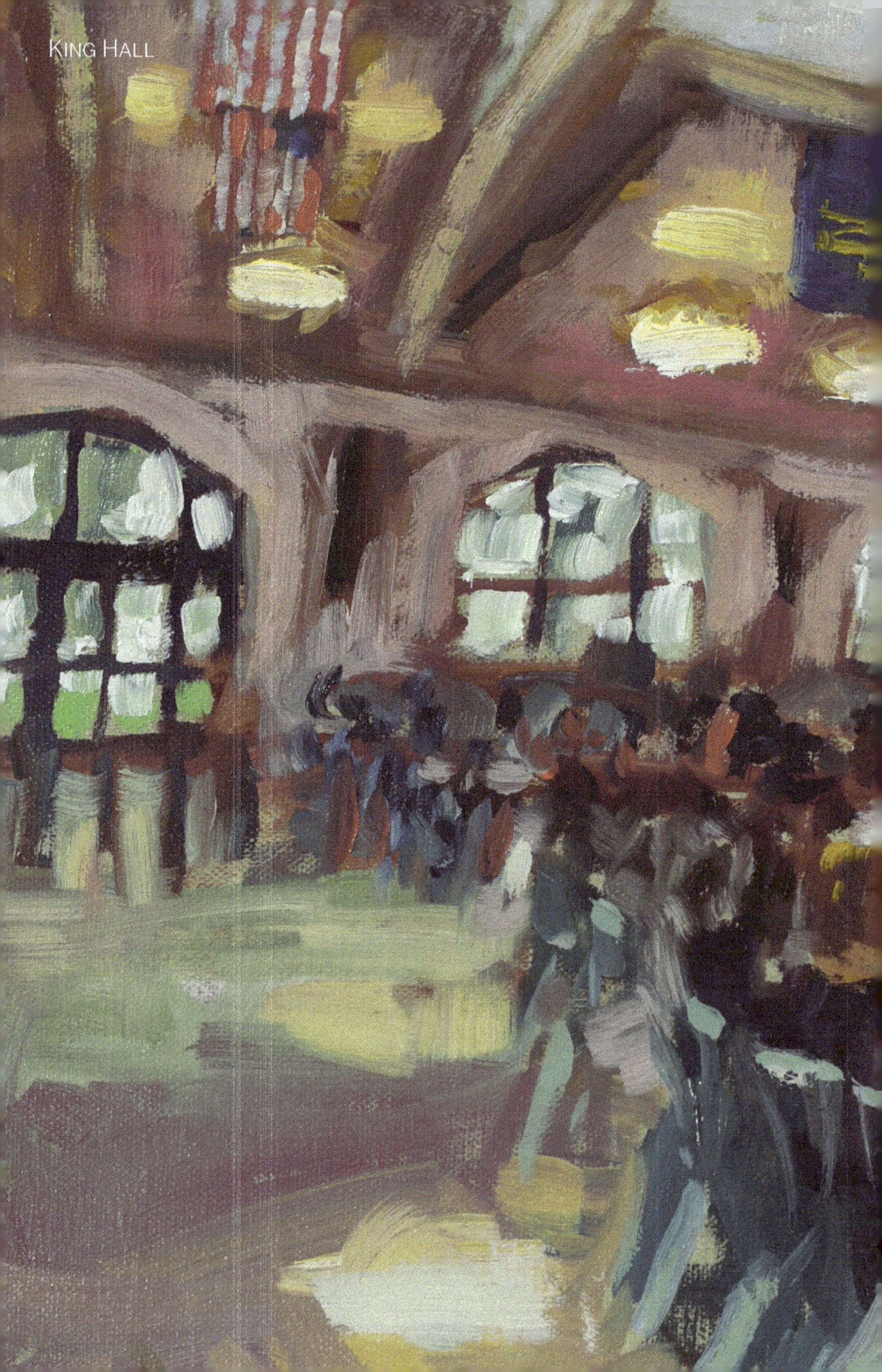

King Hall

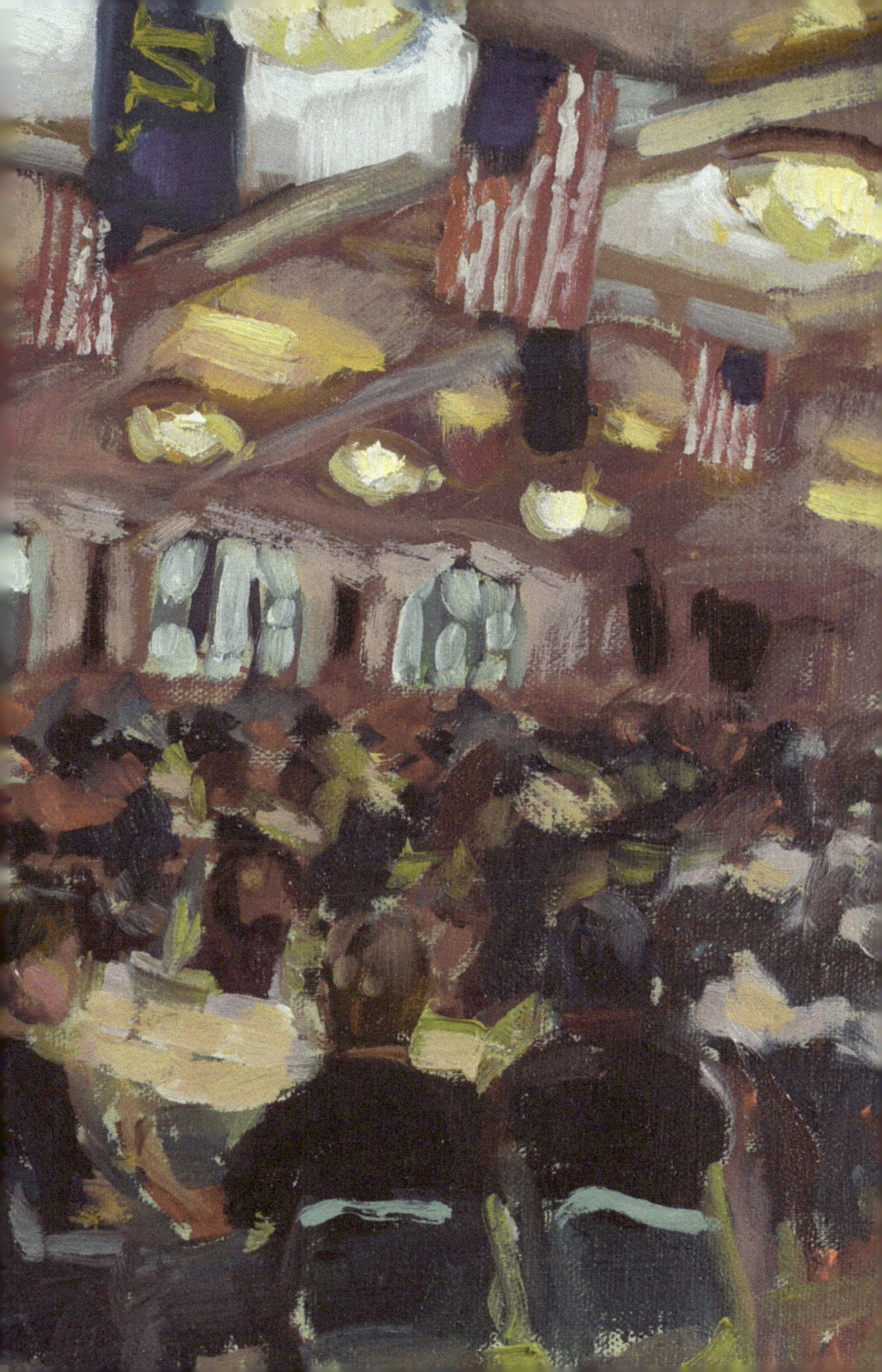

What was your favorite King Hall meal?

Did King Hall introduce you to anything you had never tried before?

Company tables or sports tables ?

Did you sit in the same spot each meal or did you roam around

the table?

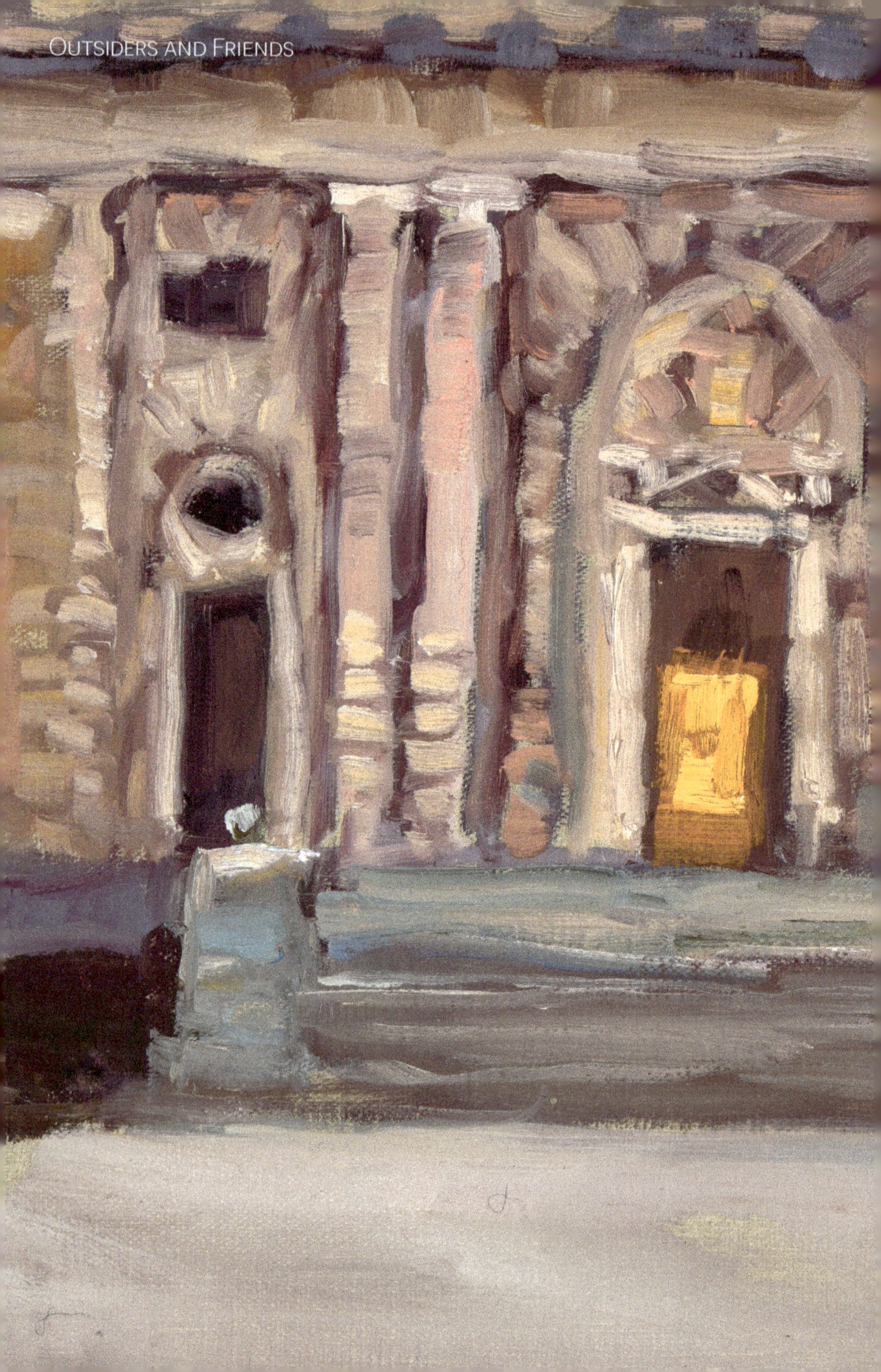
Outsiders and Friends

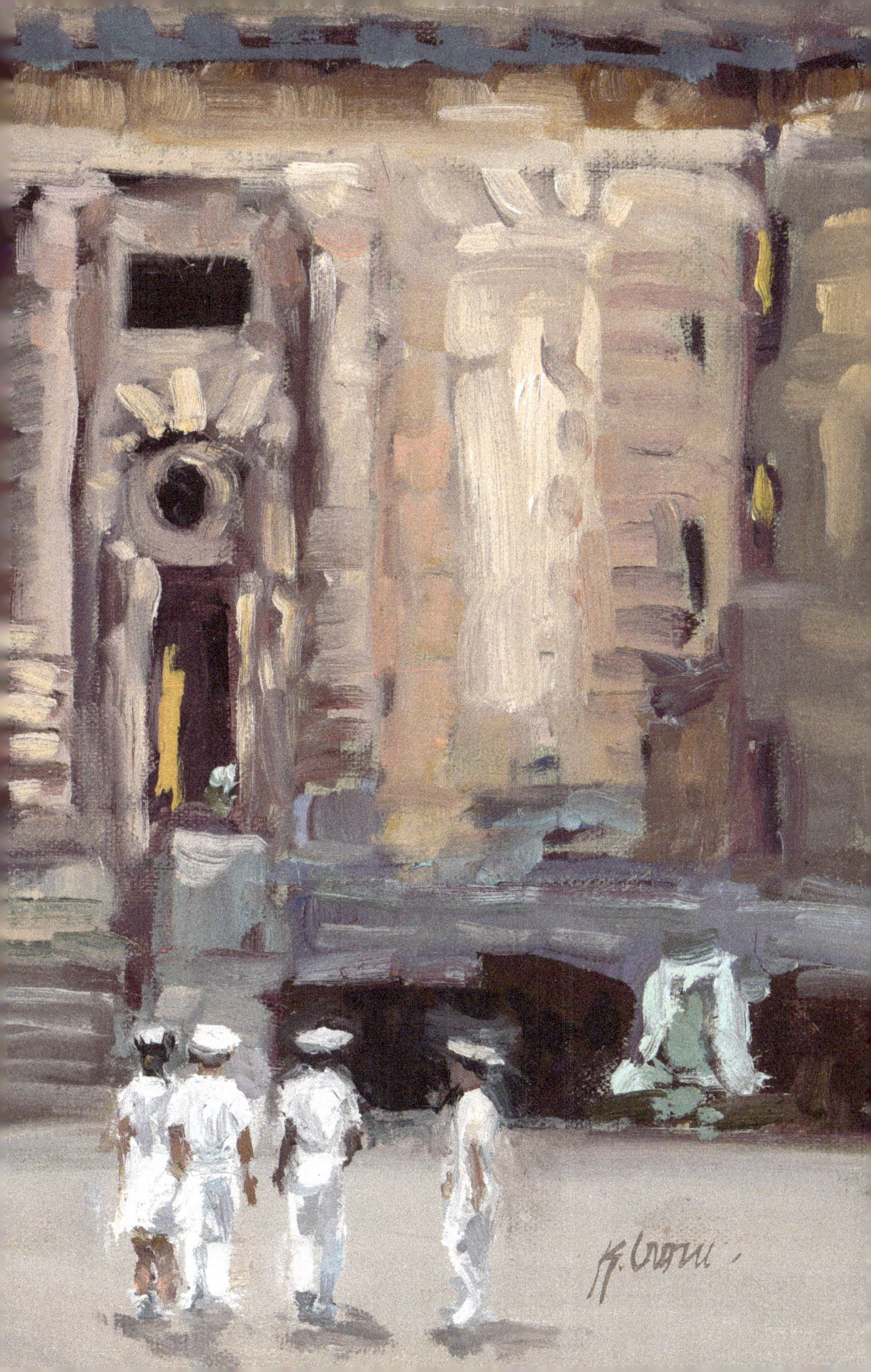

Who were your closest friends at USNA? How did you meet?

Where were your favorite spots on the Yard to relax?

Who did you room with as a Youngster? Share a memorable moment from that experience.

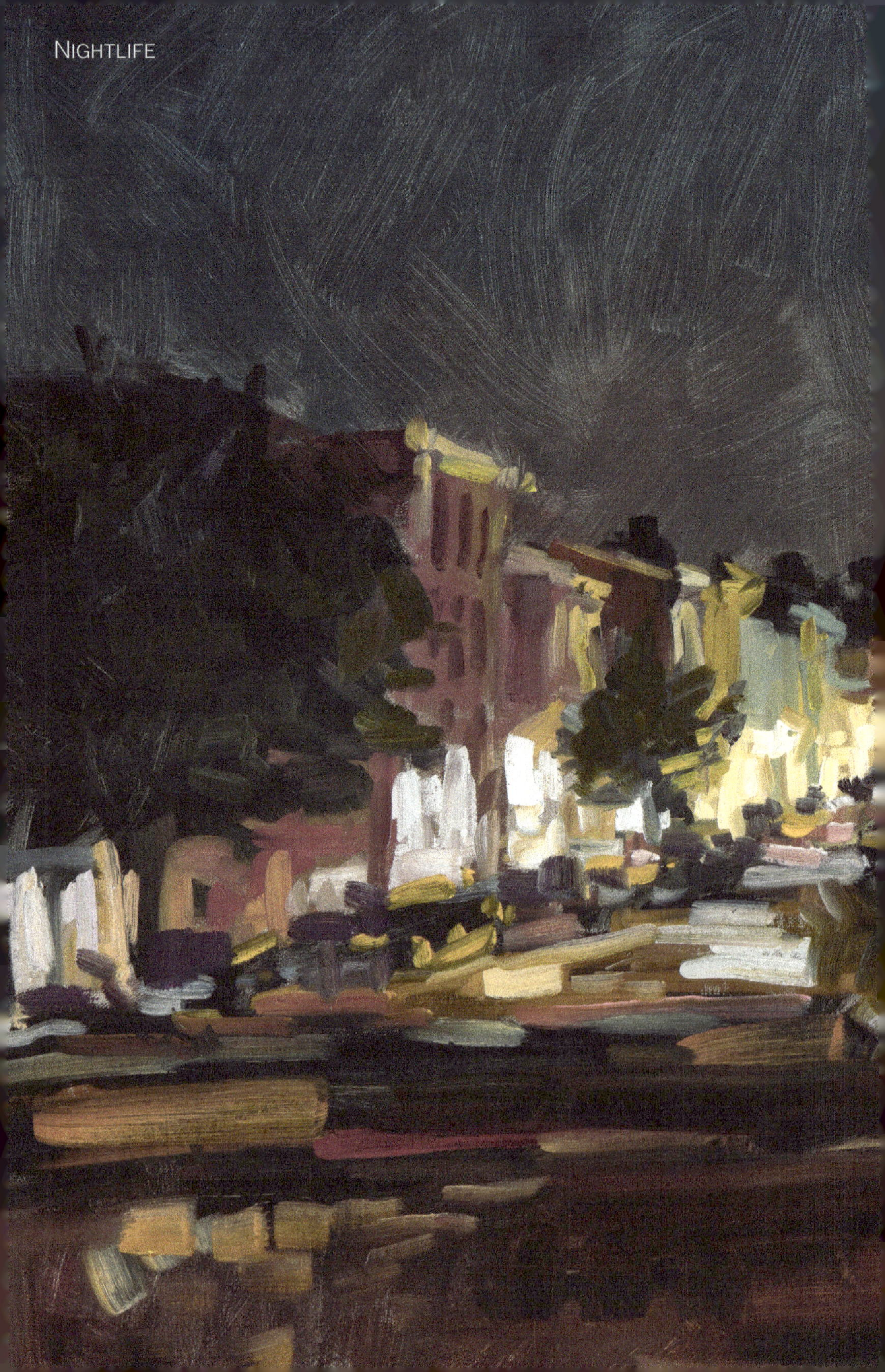
NIGHTLIFE

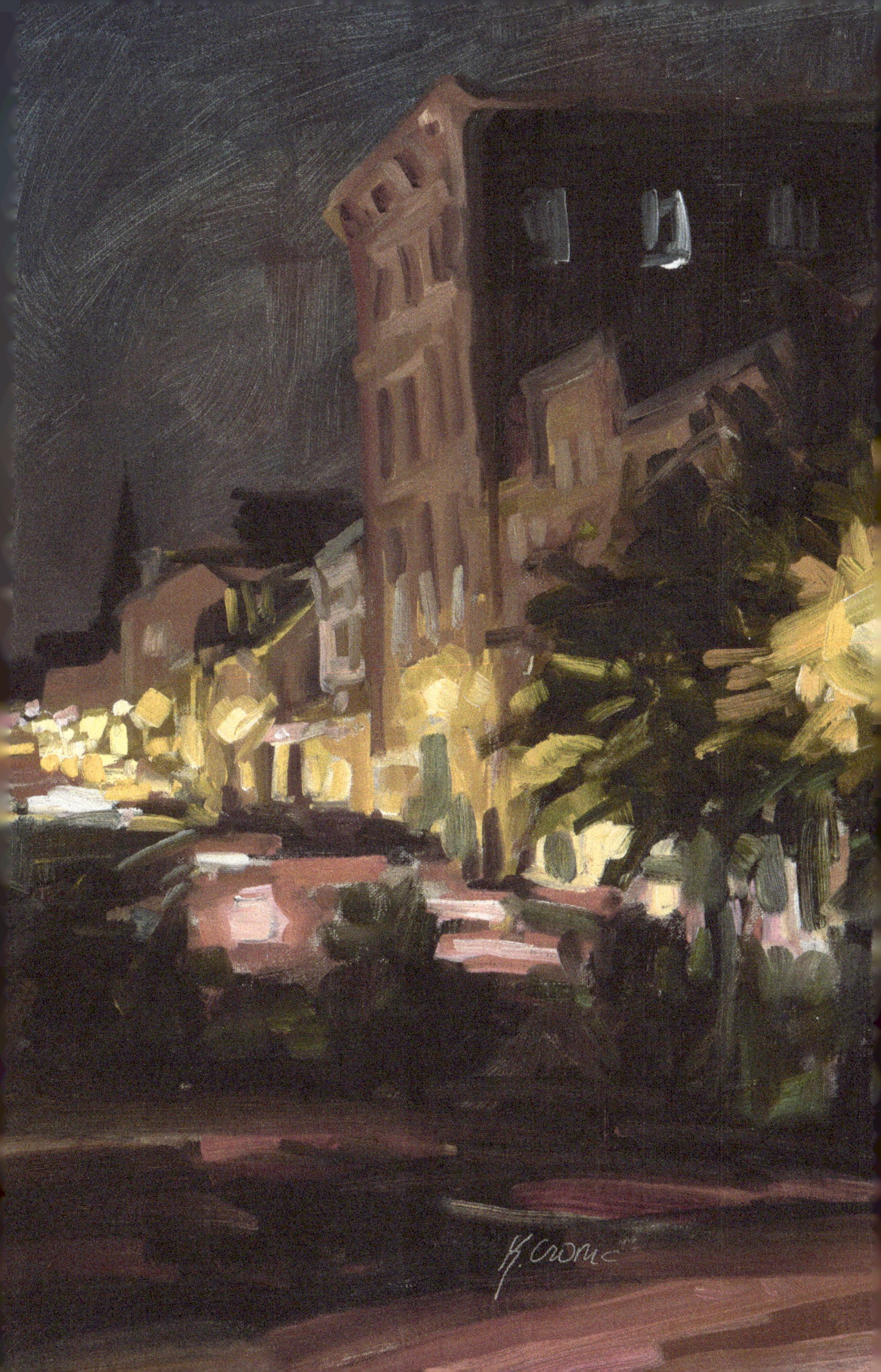

What were your favorite spots out in town?

Were you ever fried? What was your punishment?

Share a memory from a night out in DTA.

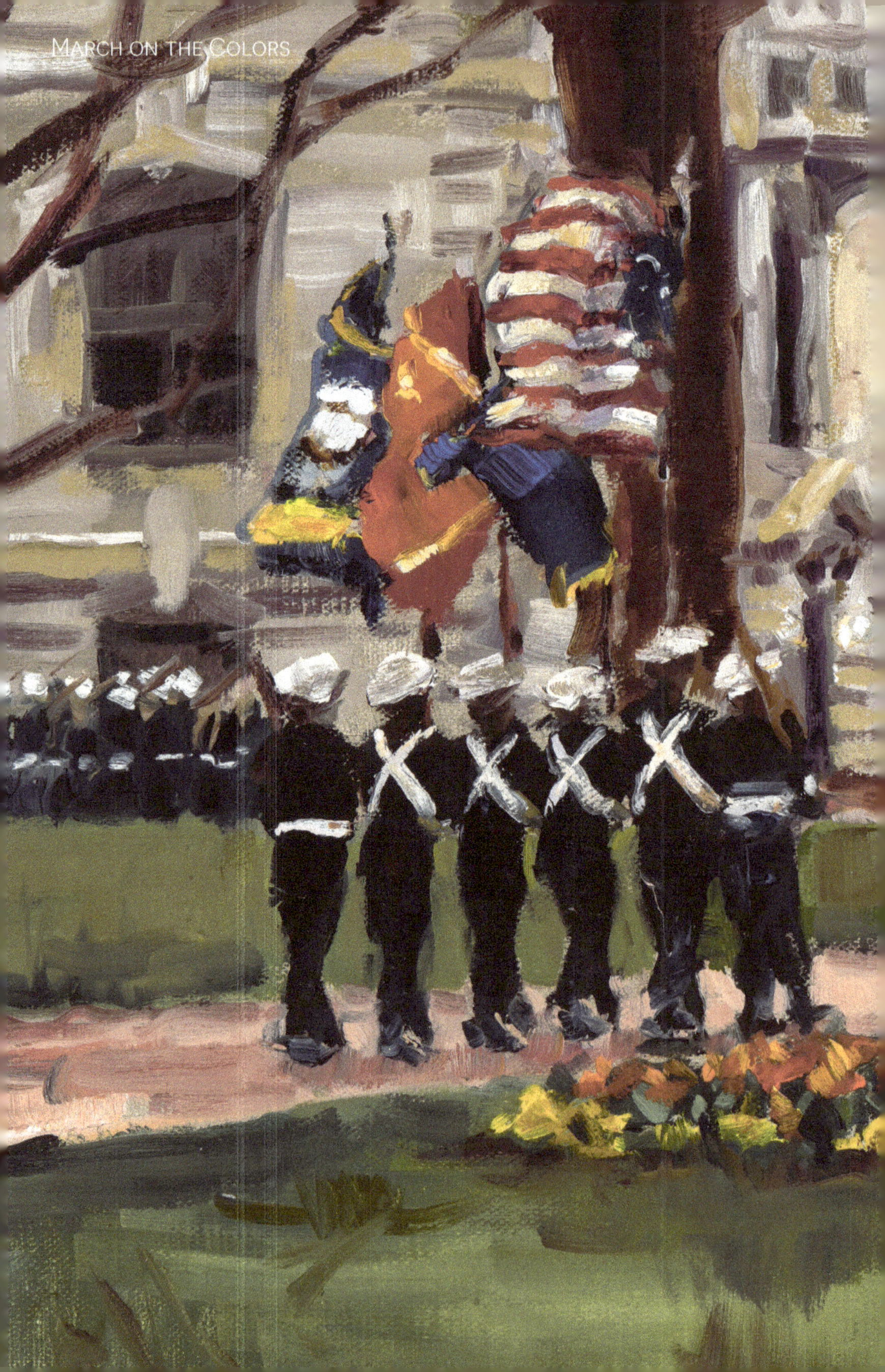
March on the Colors

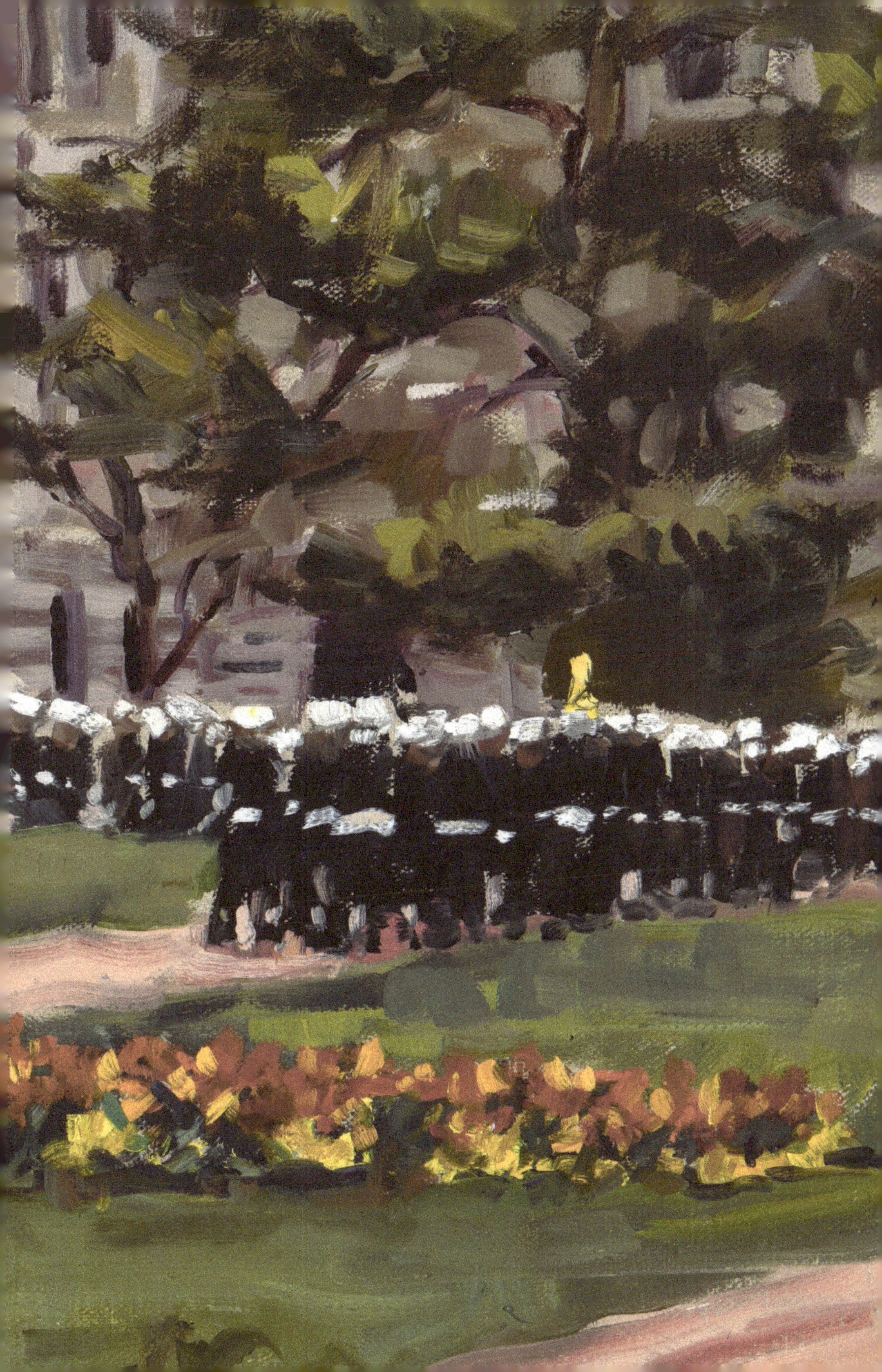

What was your favorite thing about fall on the Yard?

Share a memorable moment from marching in parades.

What new things did you try as a midshipman?

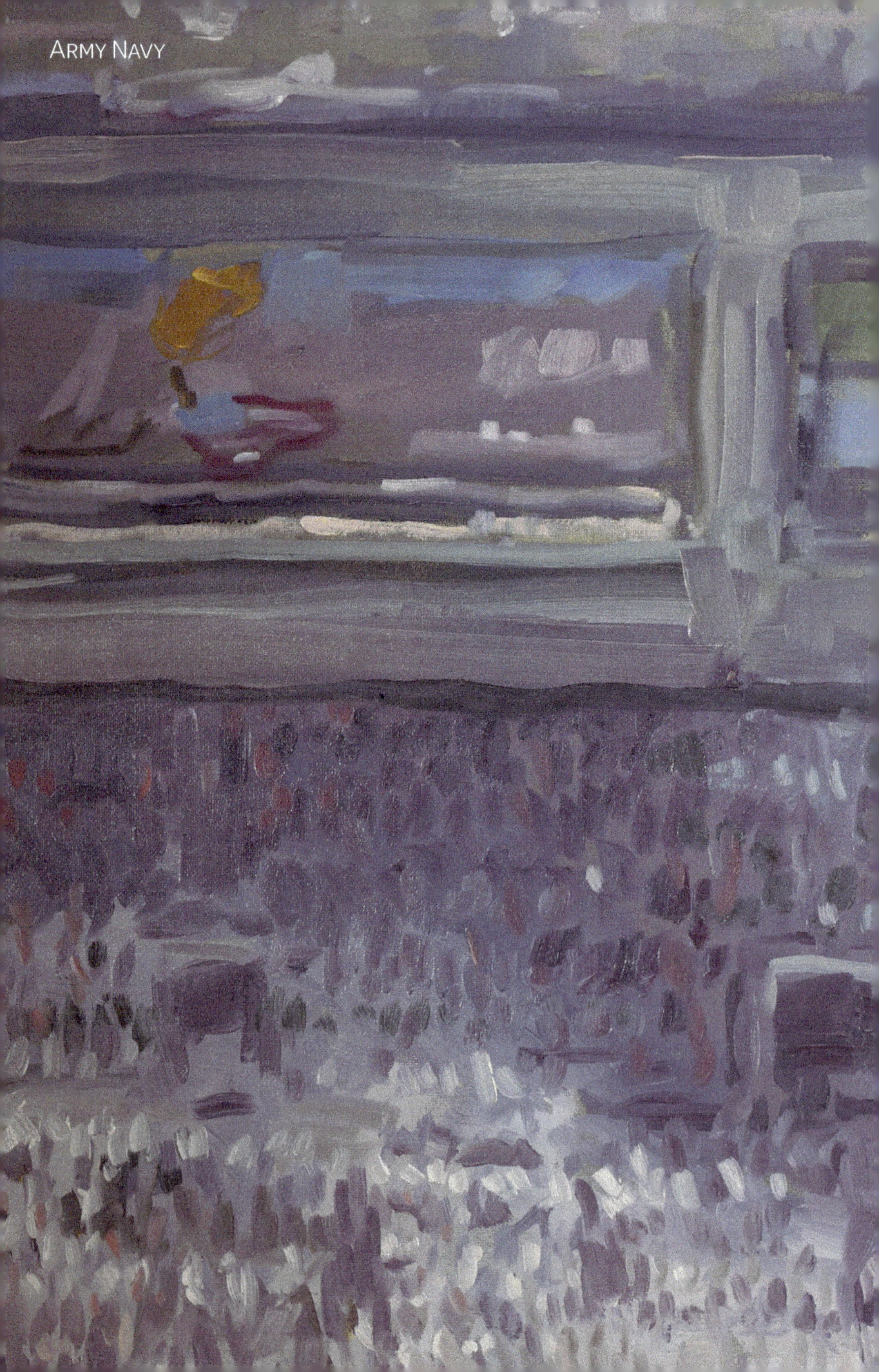
Army Navy

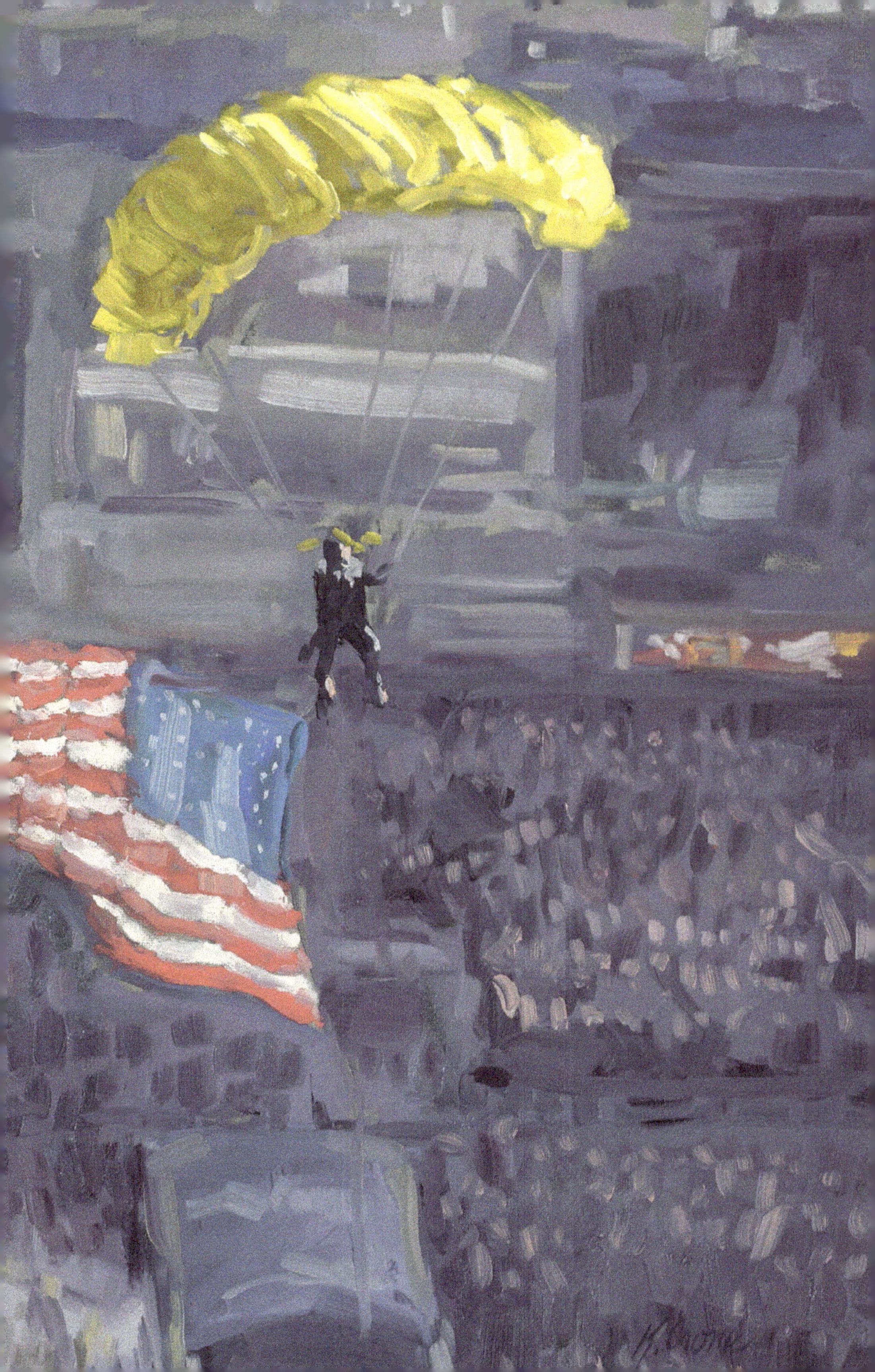

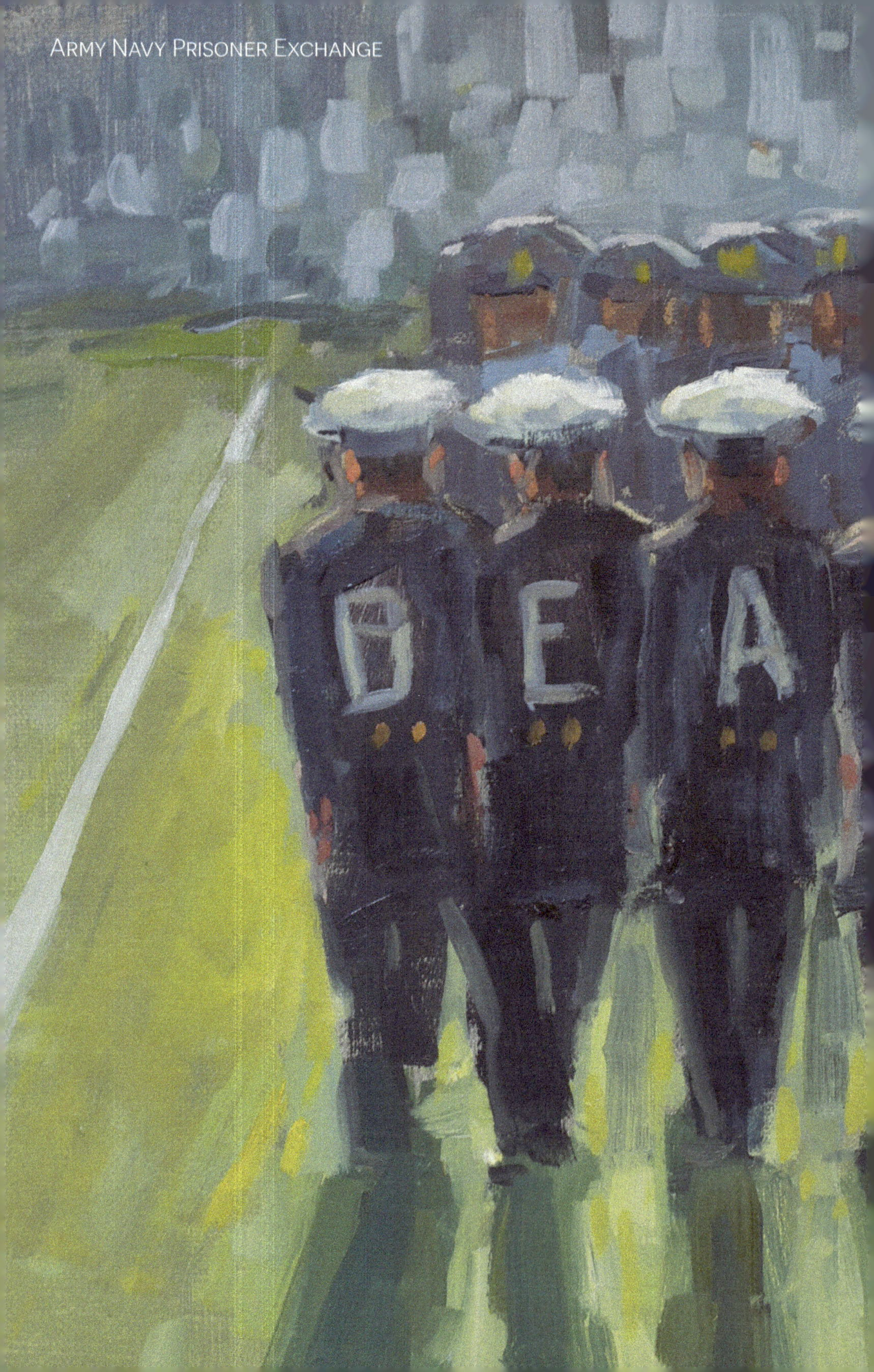

Army Navy Prisoner Exchange

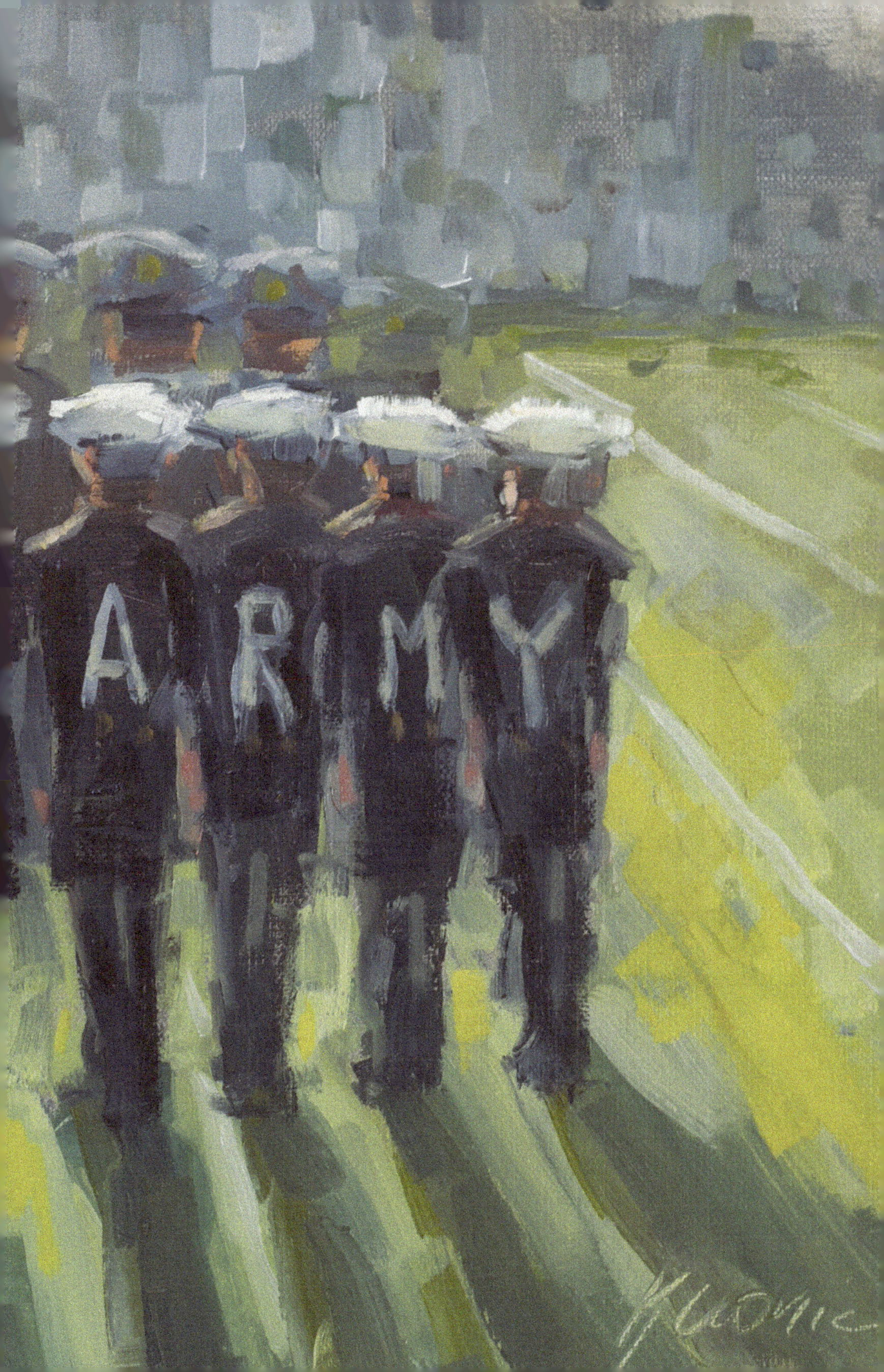

ARMY

Write down your best memory from an Army Navy game.

Who were the exchange students in your company? What do you remember about them?

Write down your favorite Army Navy prank(s).

Jumped the Wall?	Yes	No
2% Club?	Yes	No
Black N?	Yes	No
Dated a fellow mid?	Yes	No
Stepped on the seal?	Yes	No
Validated Swim Class?	Yes	No

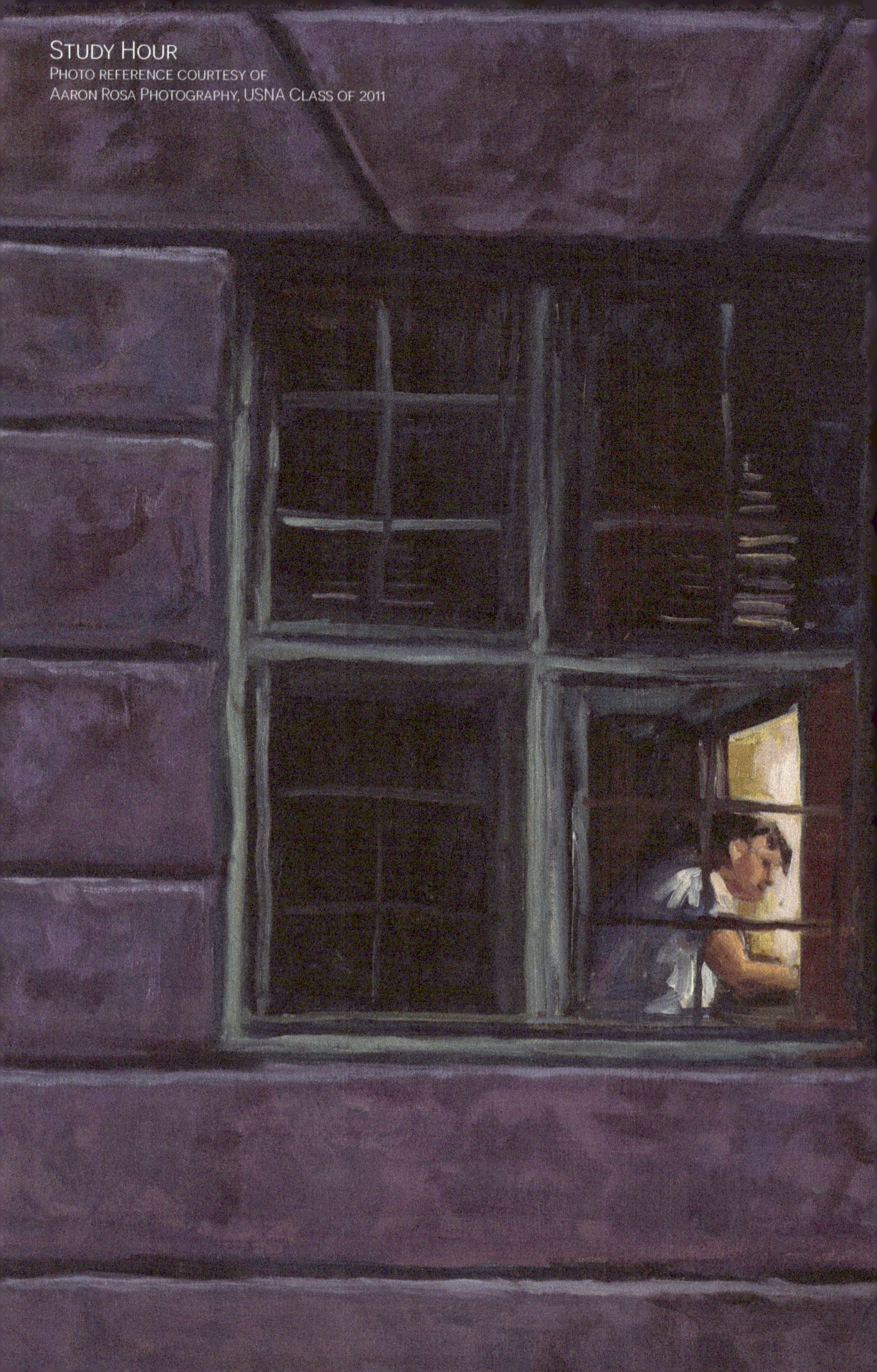

Study Hour
Photo reference courtesy of
Aaron Rosa Photography, USNA Class of 2011

What was your experience with academics at USNA?

Describe a professor that had a positive impact on you.

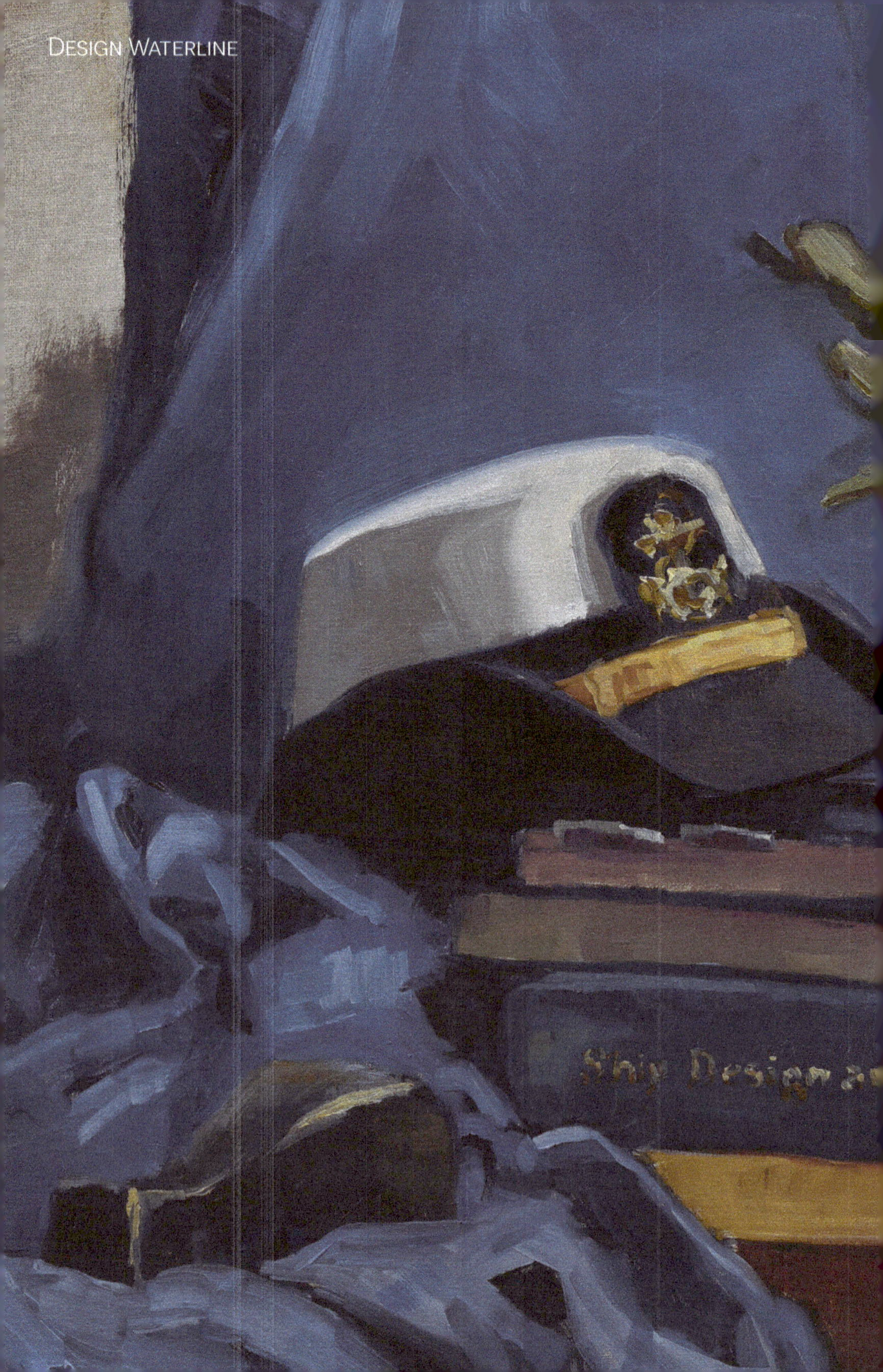
Design Waterline
Ship Design a

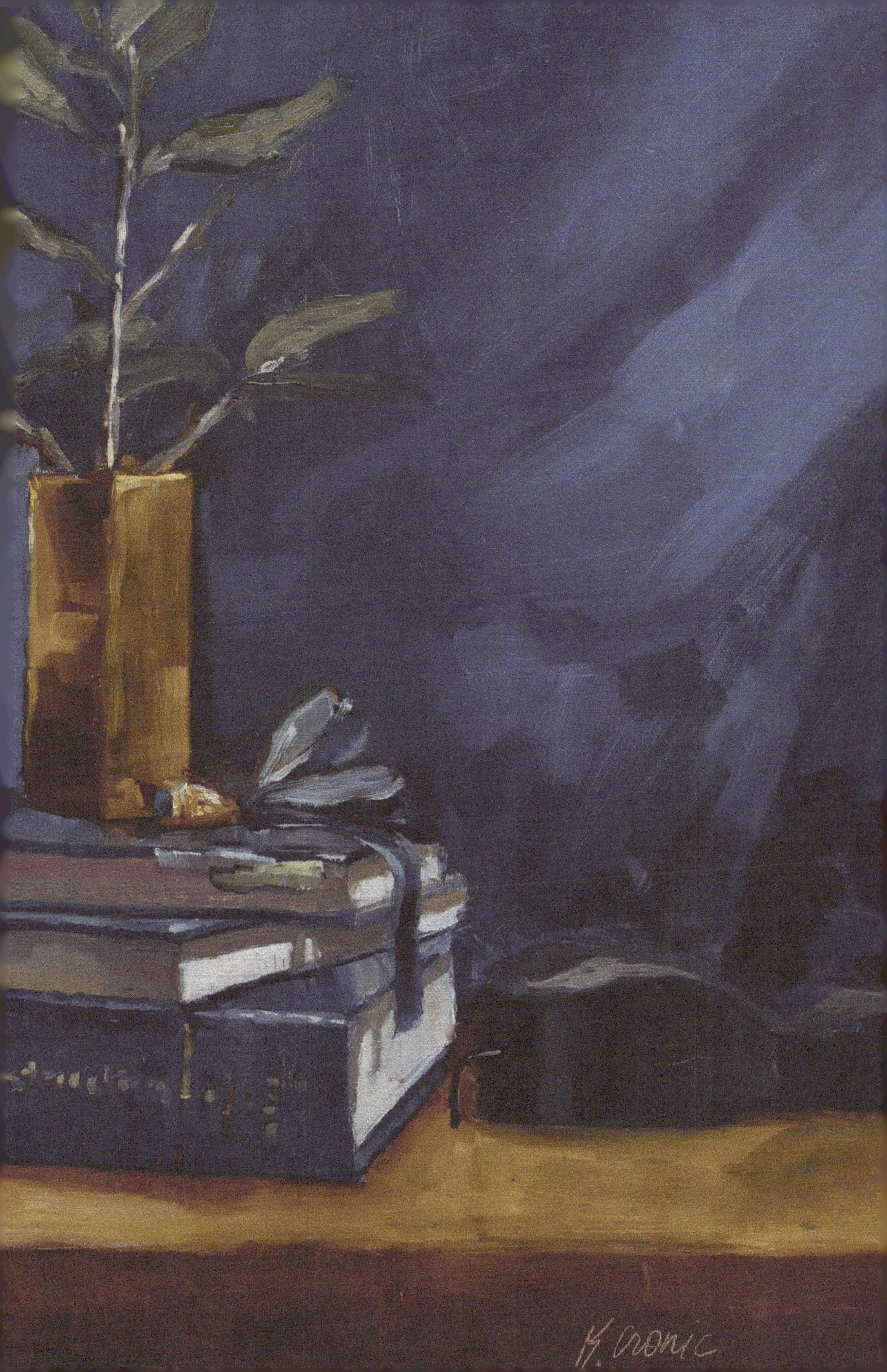
K. Cronic

Why did you choose your major? Did it change?

Where were most of your classes? What do you remember about that building?

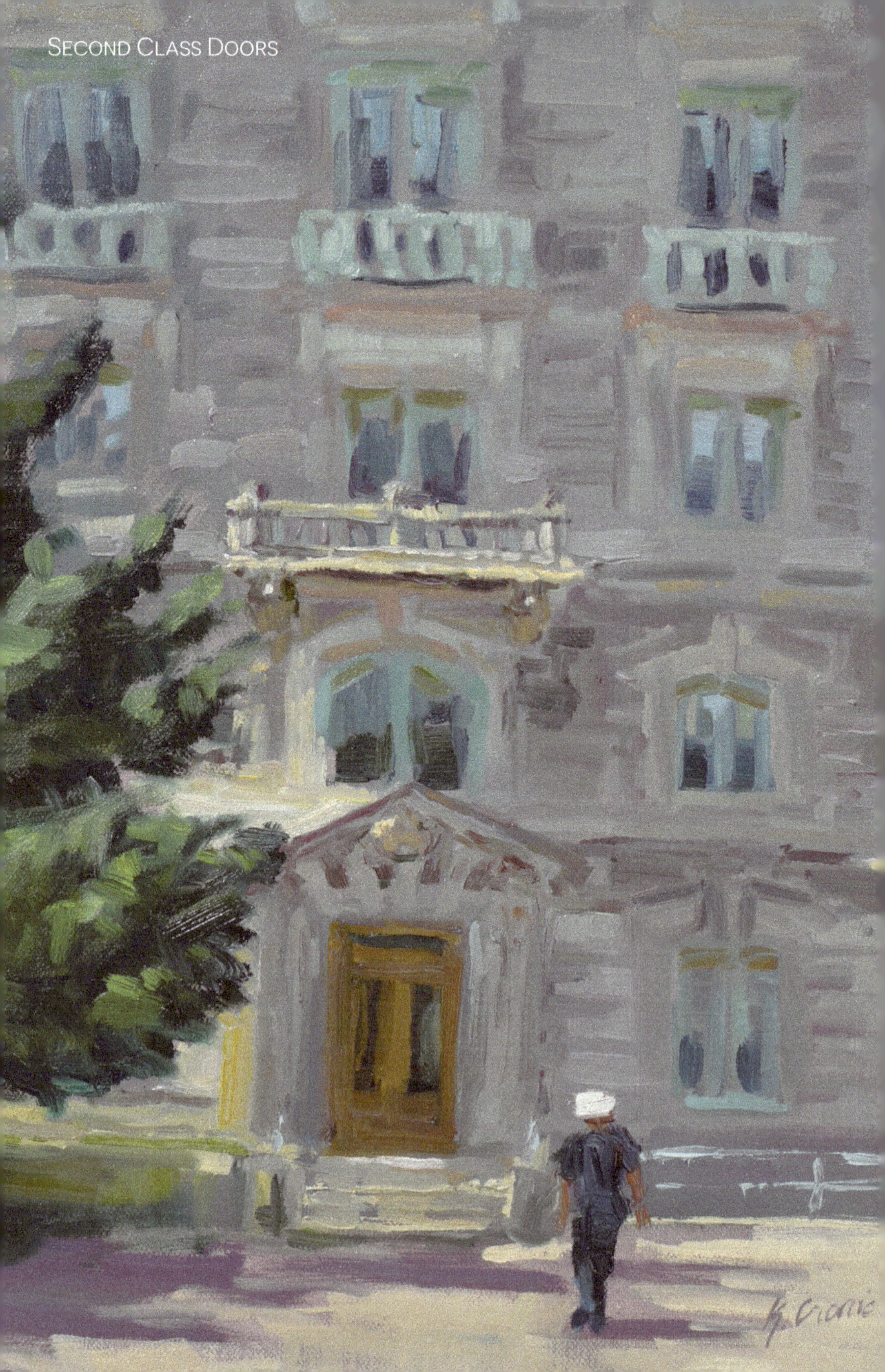
Second Class Doors

What do you remember about signing 2 for 7's? Was that a hard decision for you?

What did you do for summer training as a Second Class?

What did you find hardest about Second Class year?

What did you hope to service select as a Second Class? Did that

change from when you were a plebe?

Did you hold any billets as a Youngster or Second Class? If so, what were they and how was your experience?

Who did you room with as a Second Class? Share a memory from that experience.

THE DARK AGES

What kept you encouraged during the Dark Ages?

Did you ever have a snow day while at USNA? If so, how did you

spend it?

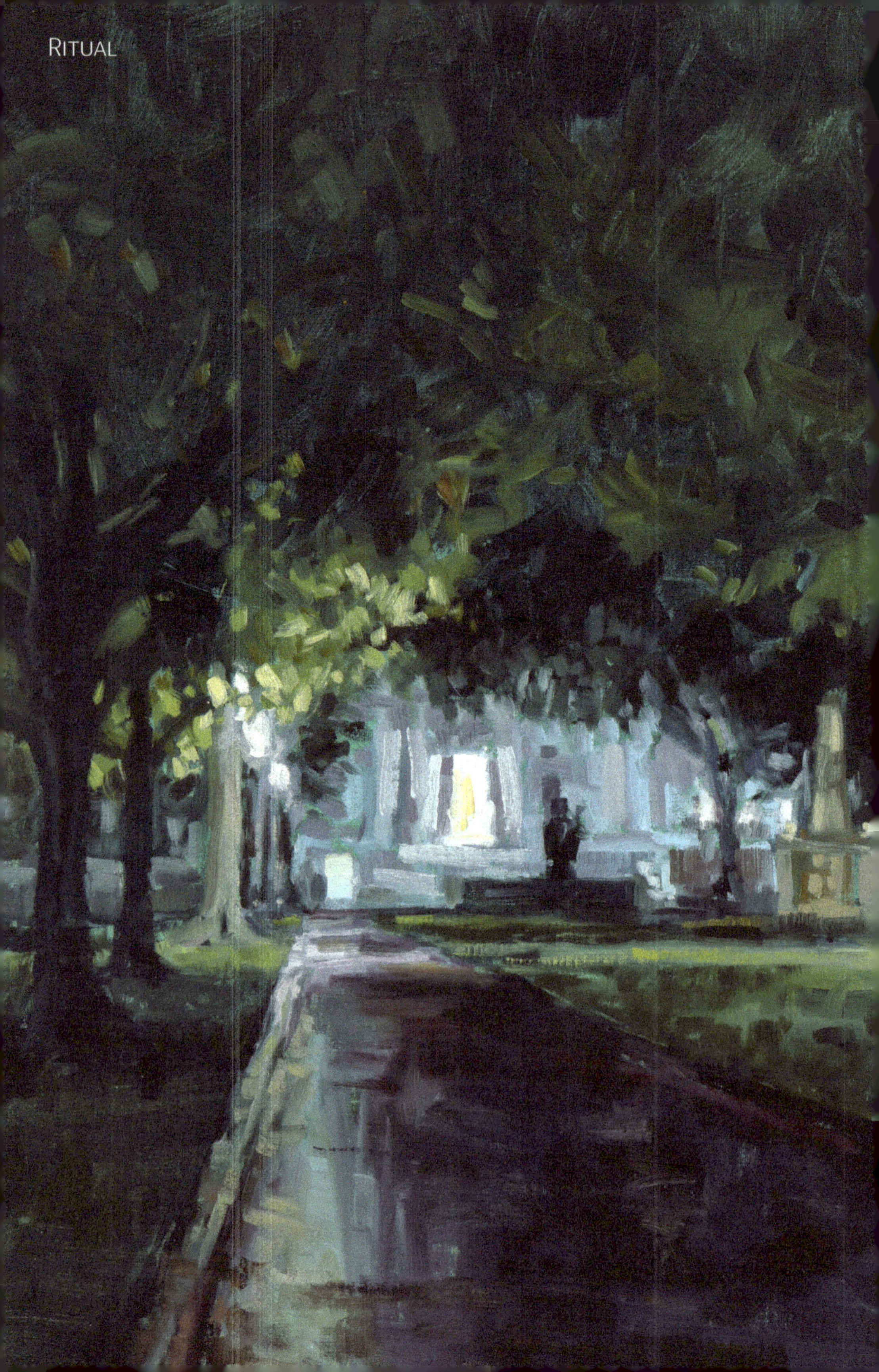
Ritual

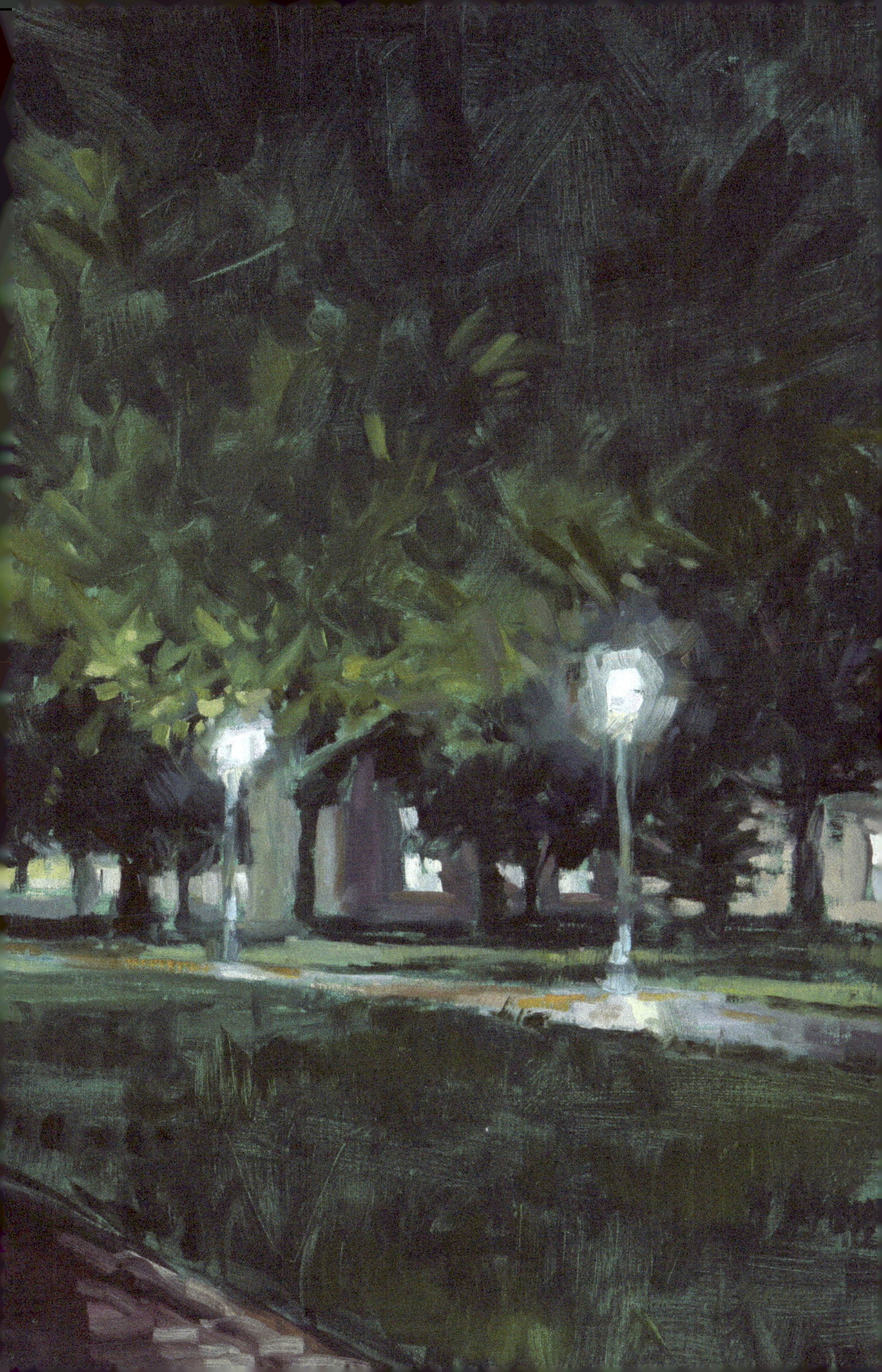

Describe an accomplishment from USNA that you are proud of.

Rickover or Nimitz?

Inner or Outer?

Ricketts // Wesley Brown // 7th Wing // Halsey // Mac D

Varsity // Club // Intramurals

Naps or Striper Sword Practice?

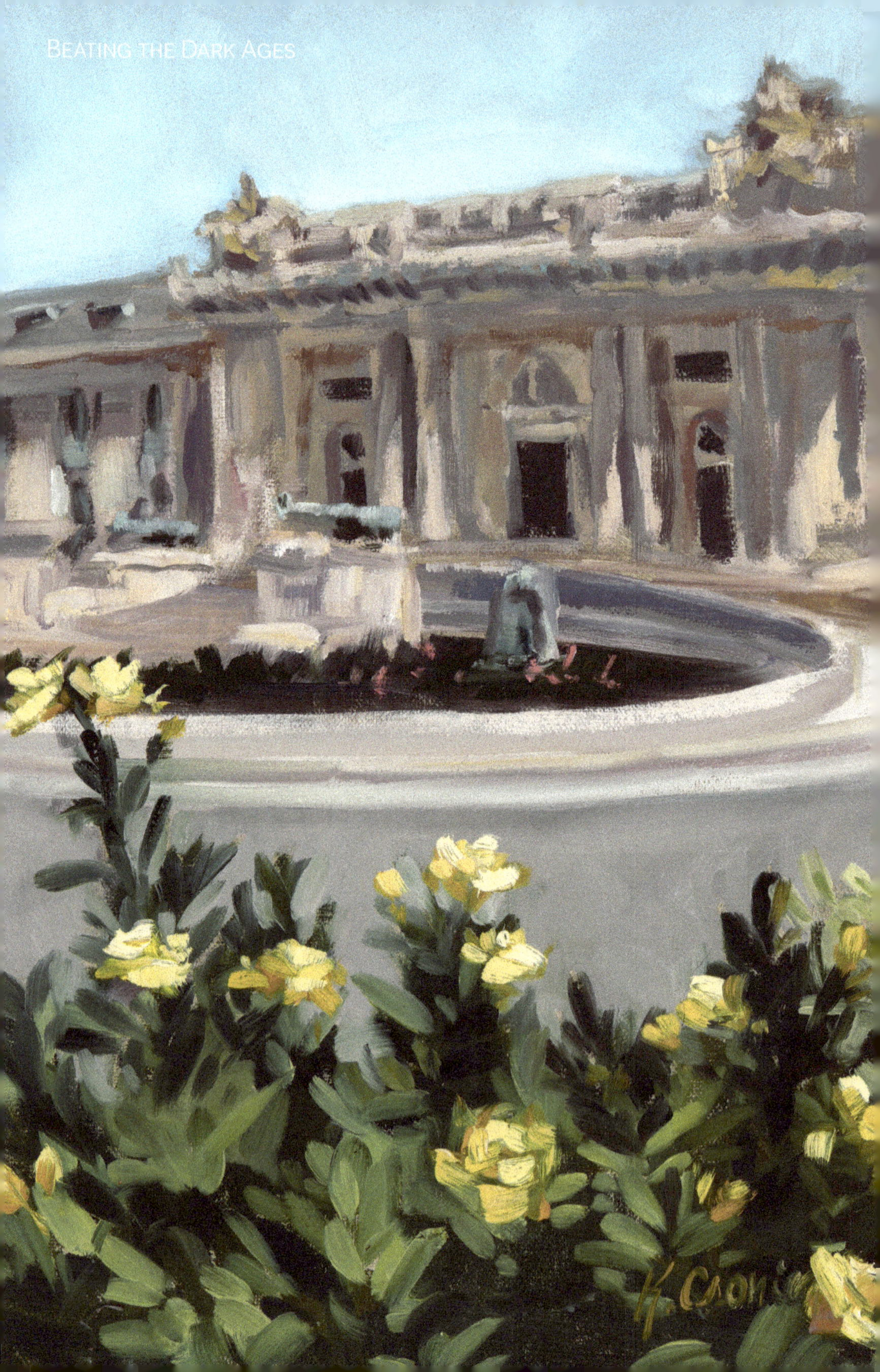
Beating the Dark Ages

What was your favorite part of spring in Annapolis?

Did you ever travel with summer training, a club, academics, or sports during your time at USNA? Write about your experience.

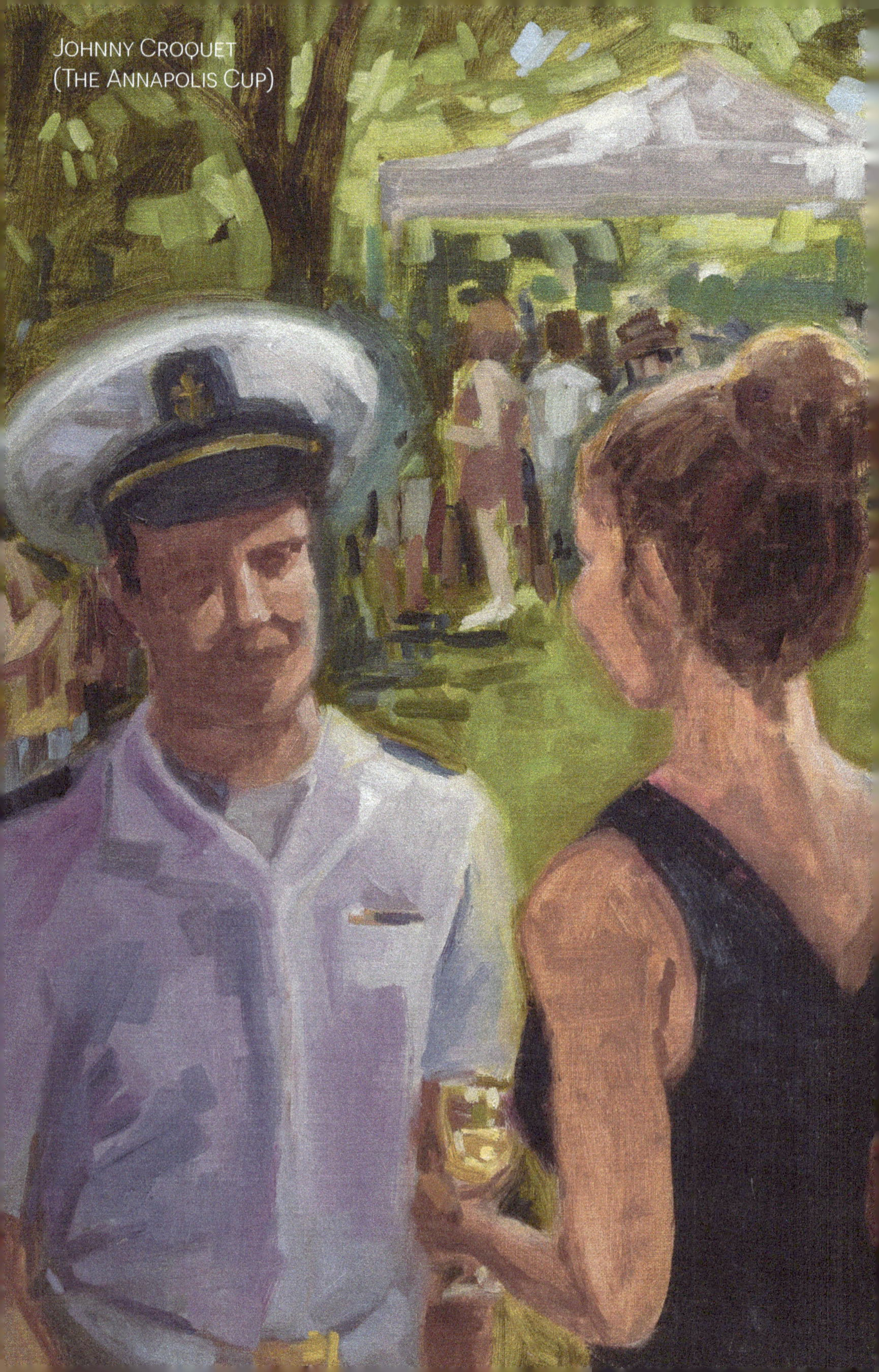

Johnny Croquet
(The Annapolis Cup)

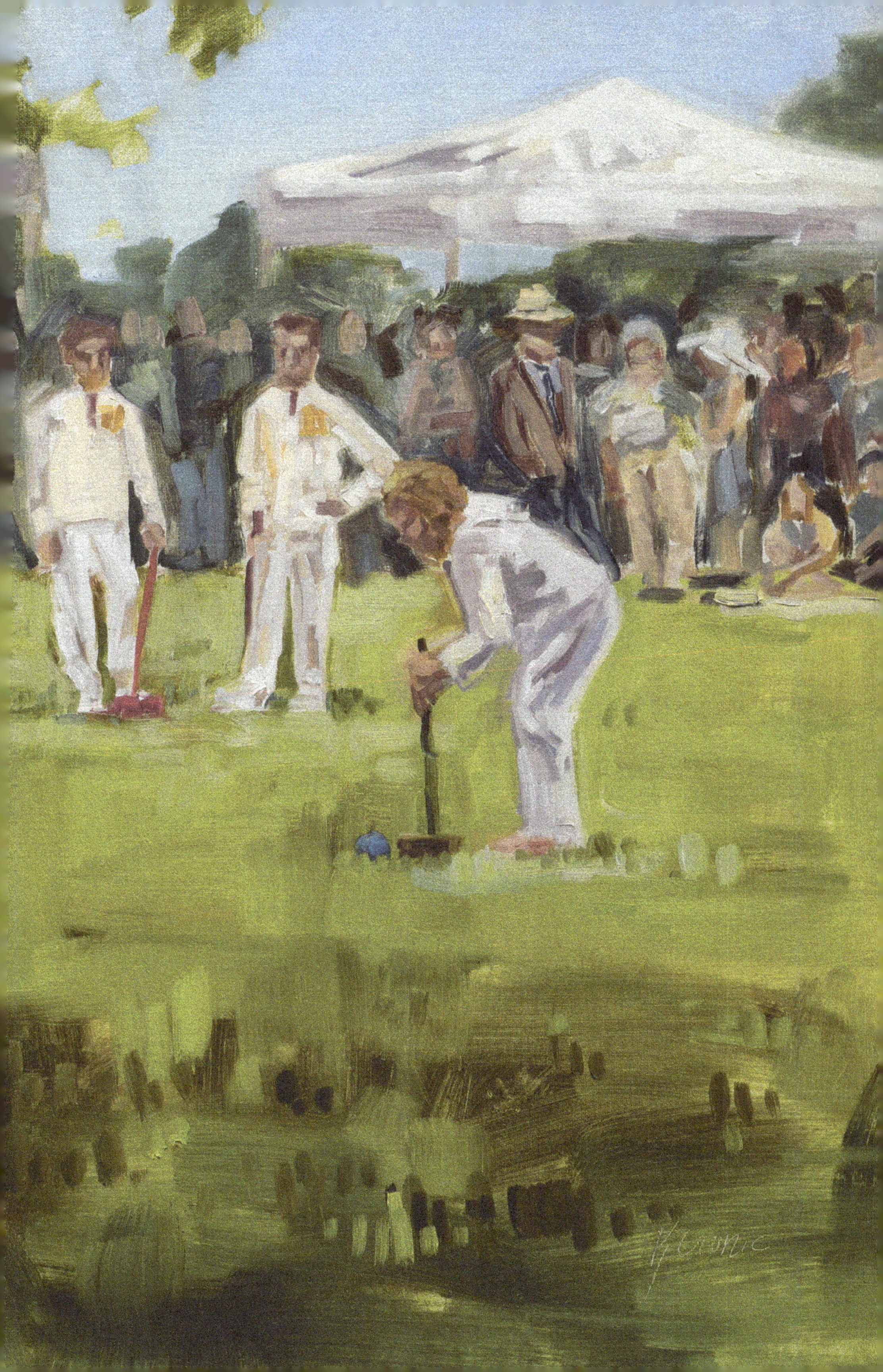

Share your favorite memory from the annual croquet tourna-

ment.

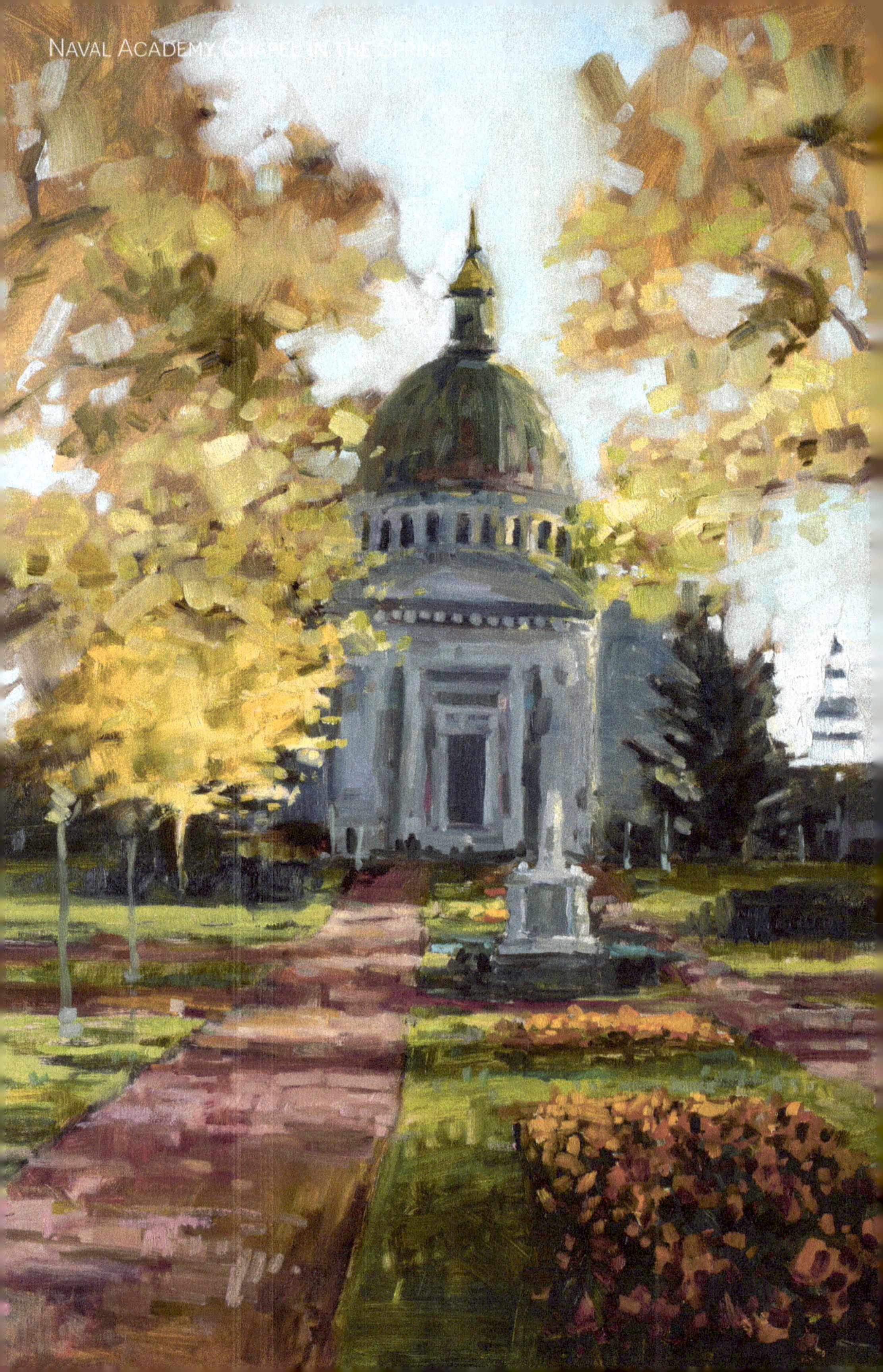

Naval Academy Chapel in the Spring

Were you involved in any religious clubs or services, on or off the Yard? Which ones, and who did you share them with? Write down any additional fond memories from your experience.

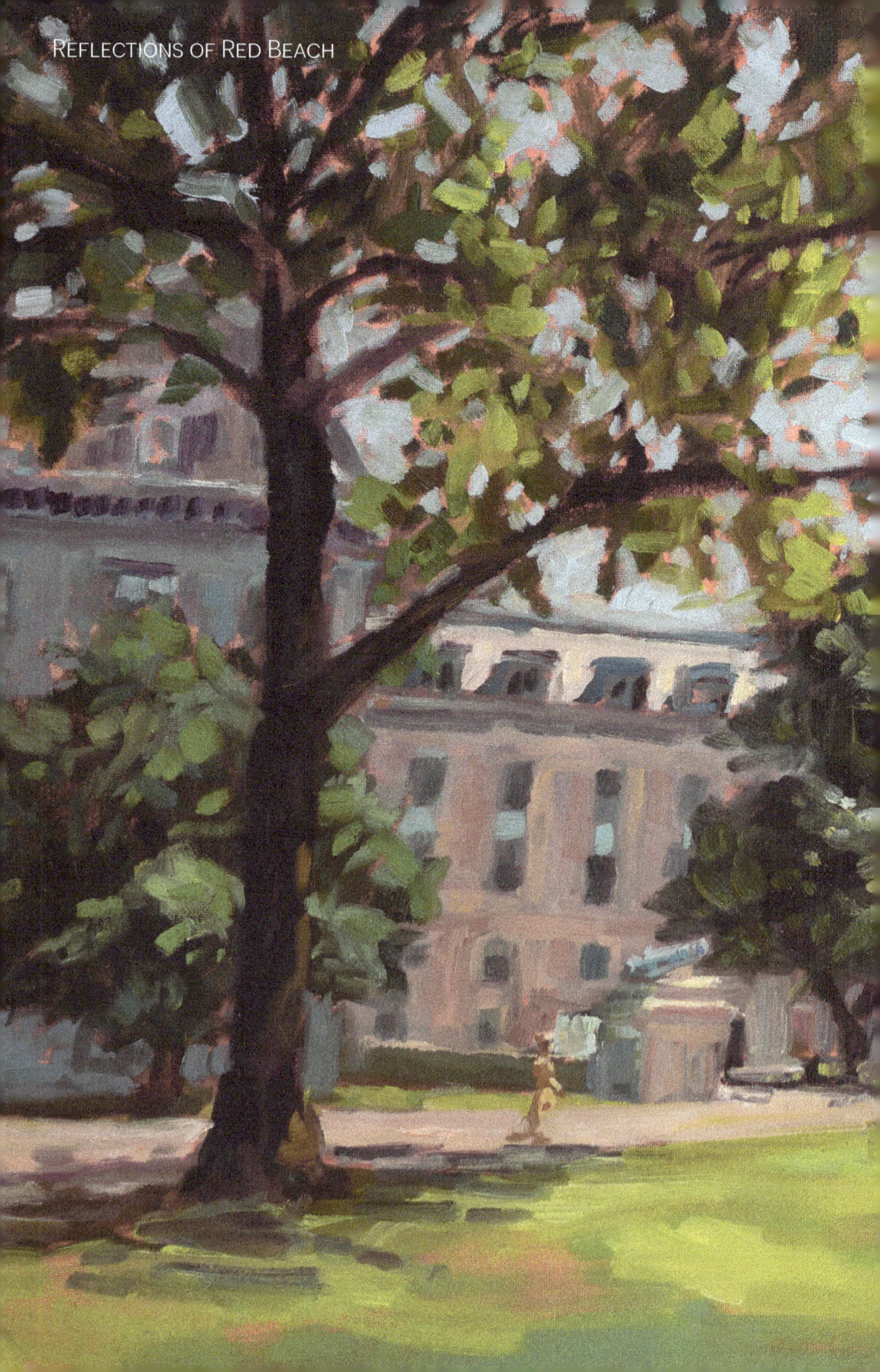
REFLECTIONS OF RED BEACH

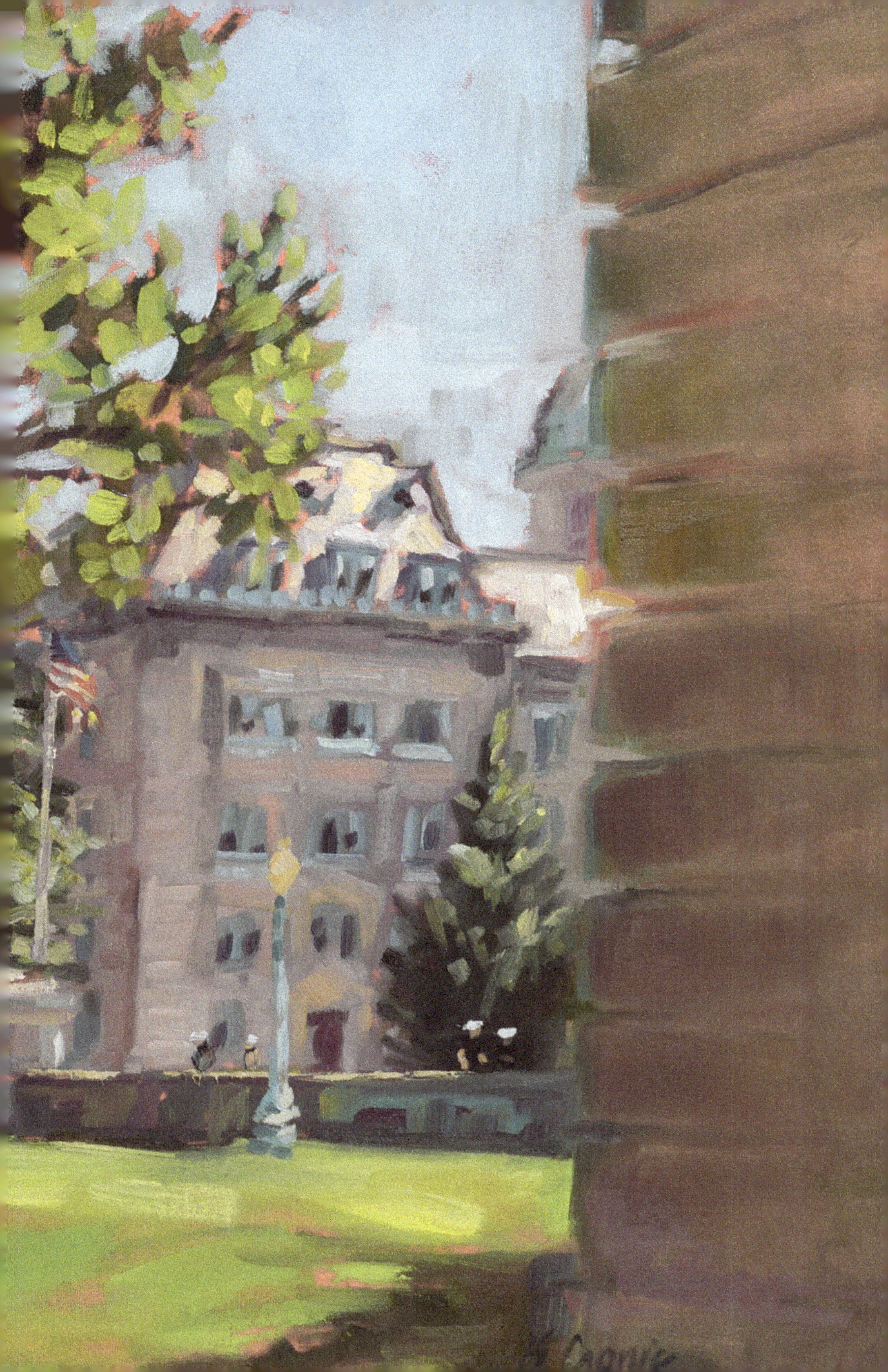

What do you remember about living in Bancroft Hall?

What was your company? What floors did you live on?

Did you ever keep animals or contraband in your room?

What is the most interesting space you discovered in the build-

ings on the Yard?

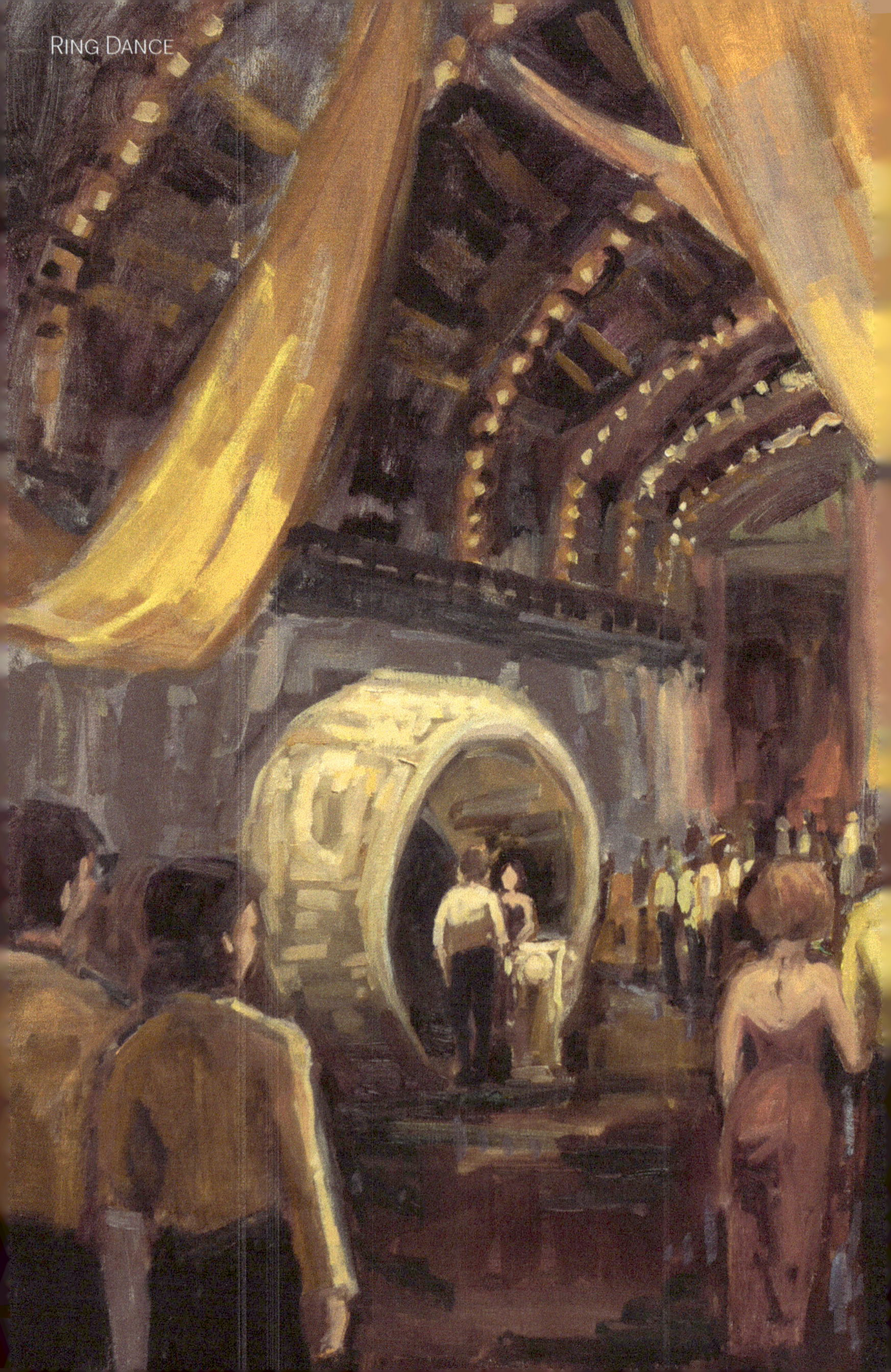
RING DANCE

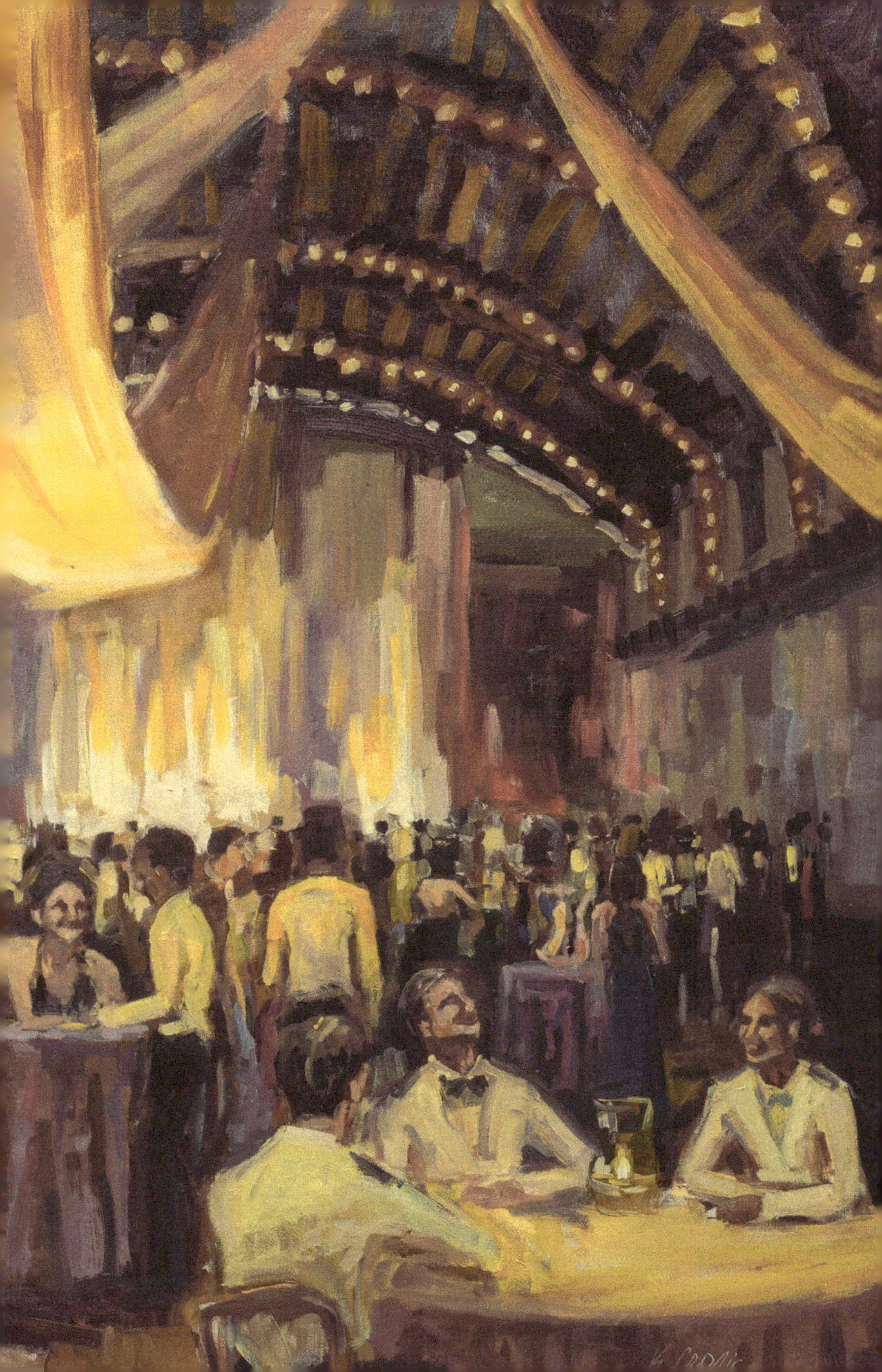

Who did you bring as your date to Ring Dance?

What was your favorite part of Ring Dance?

How did you choose your ring? Does it include any personal details special to you?

How did you feel about wearing your ring? Did that change over time?

Have you ever connected with someone outside of Annapolis

because one of you was wearing your ring?

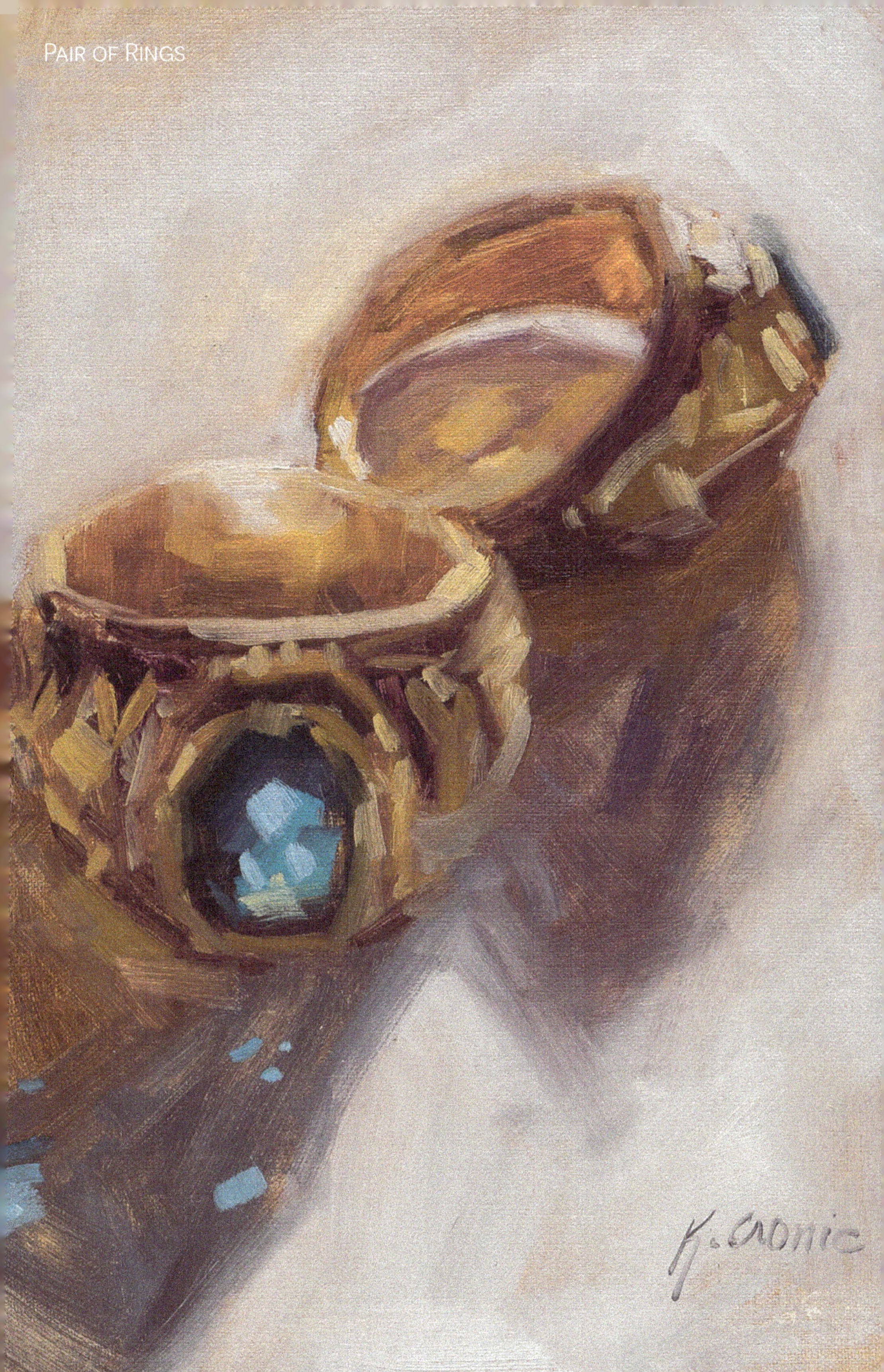
Pair of Rings

What did you do for summer training during Firstie Summer?

Name your favorite and least favorite uniforms to wear as a
midshipman.

Did you ever get away with breaking a rule? Which one(s)?

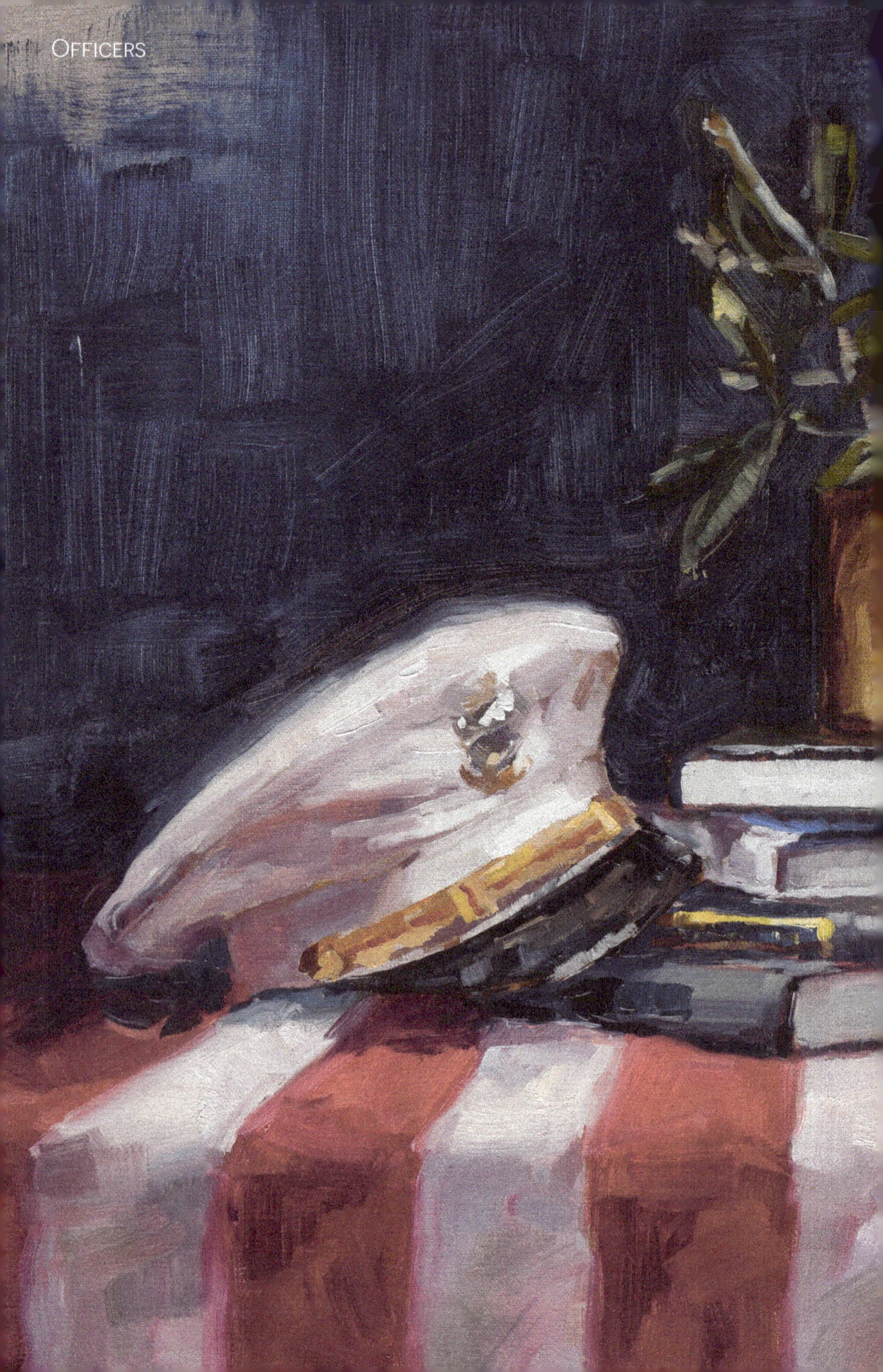

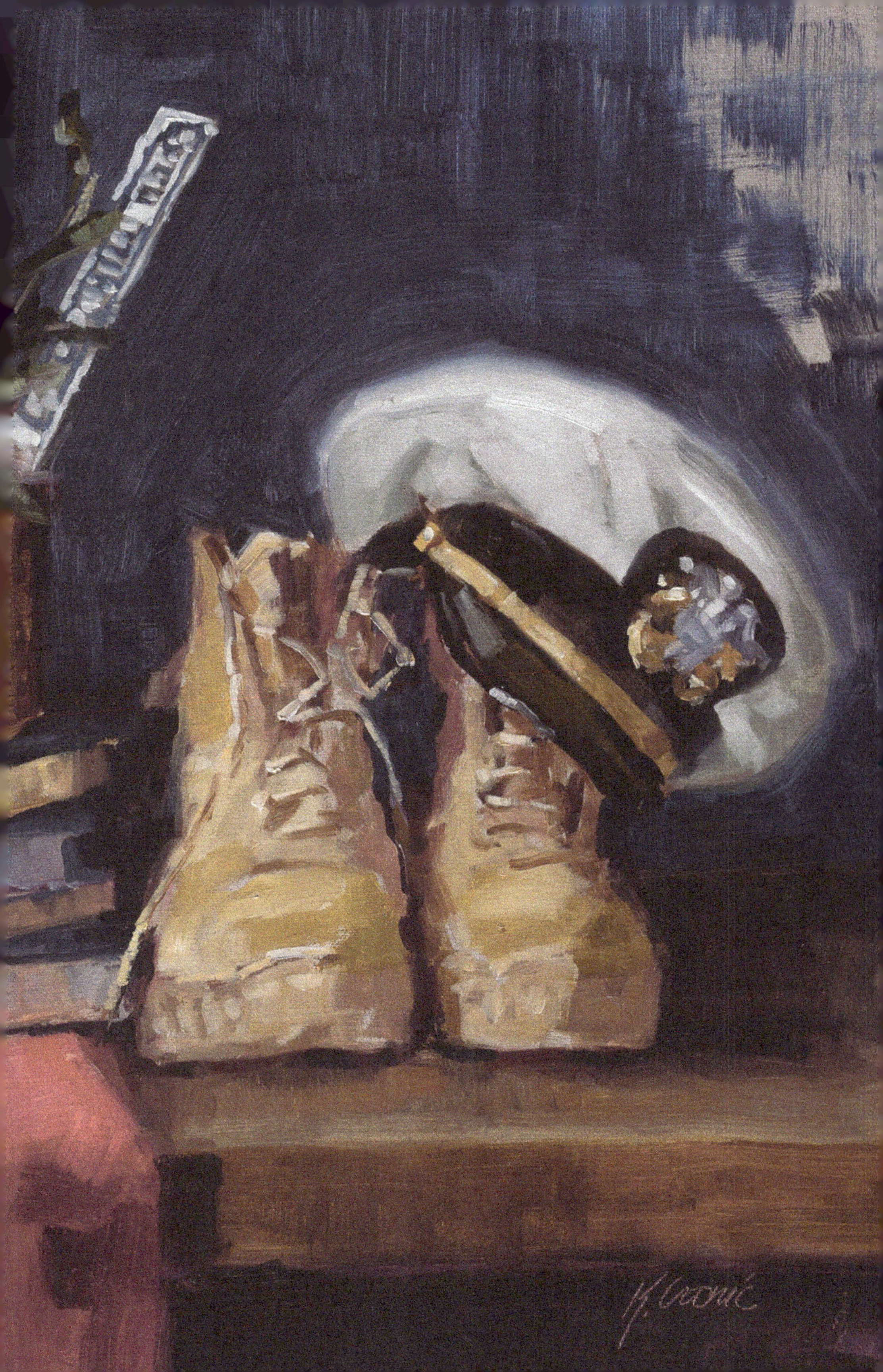

How did you come to service select what you did? Did your preferences change throughout your time? If so, why?

Share your memories of Service Selection day.

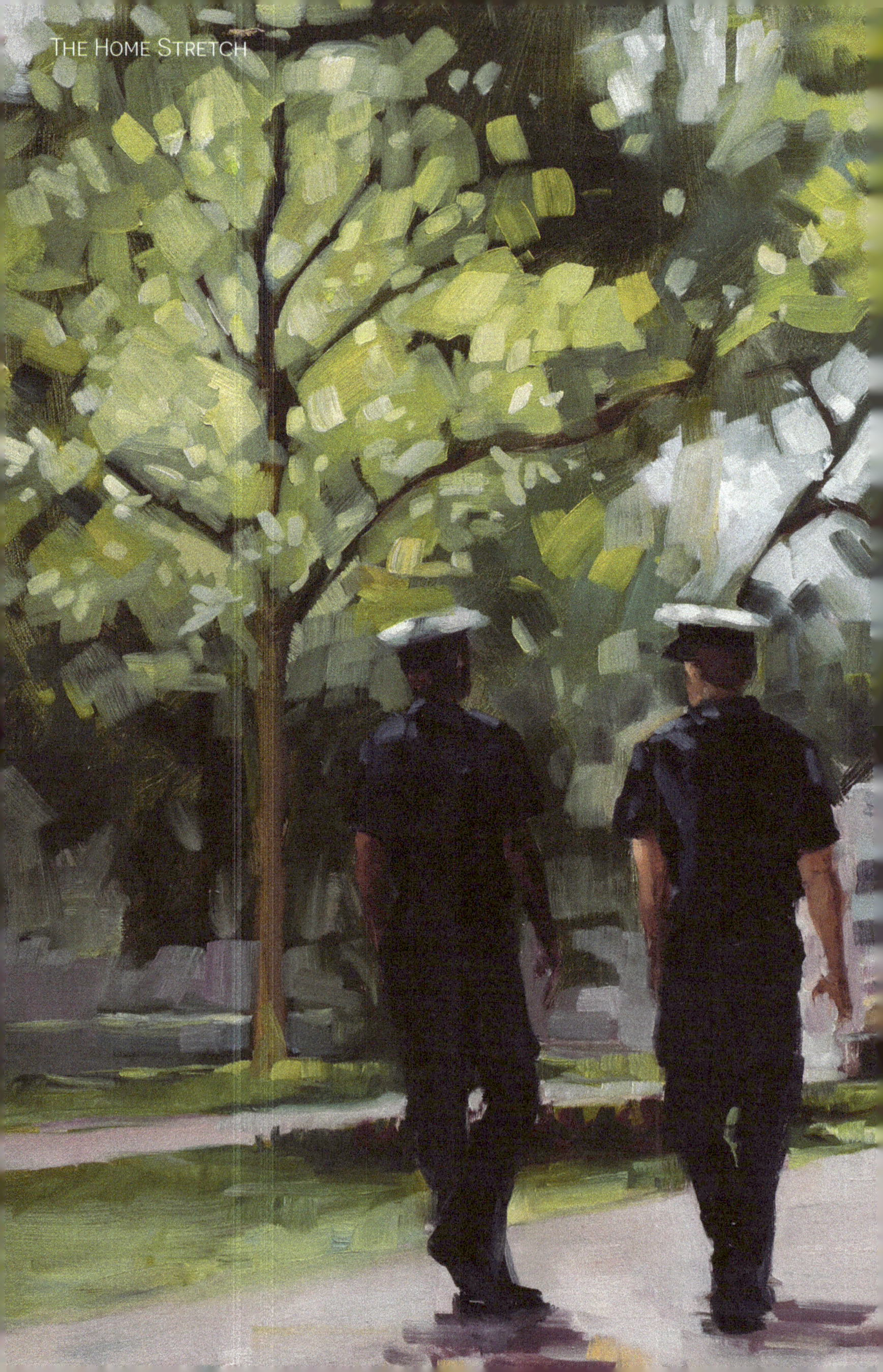
THE HOME STRETCH

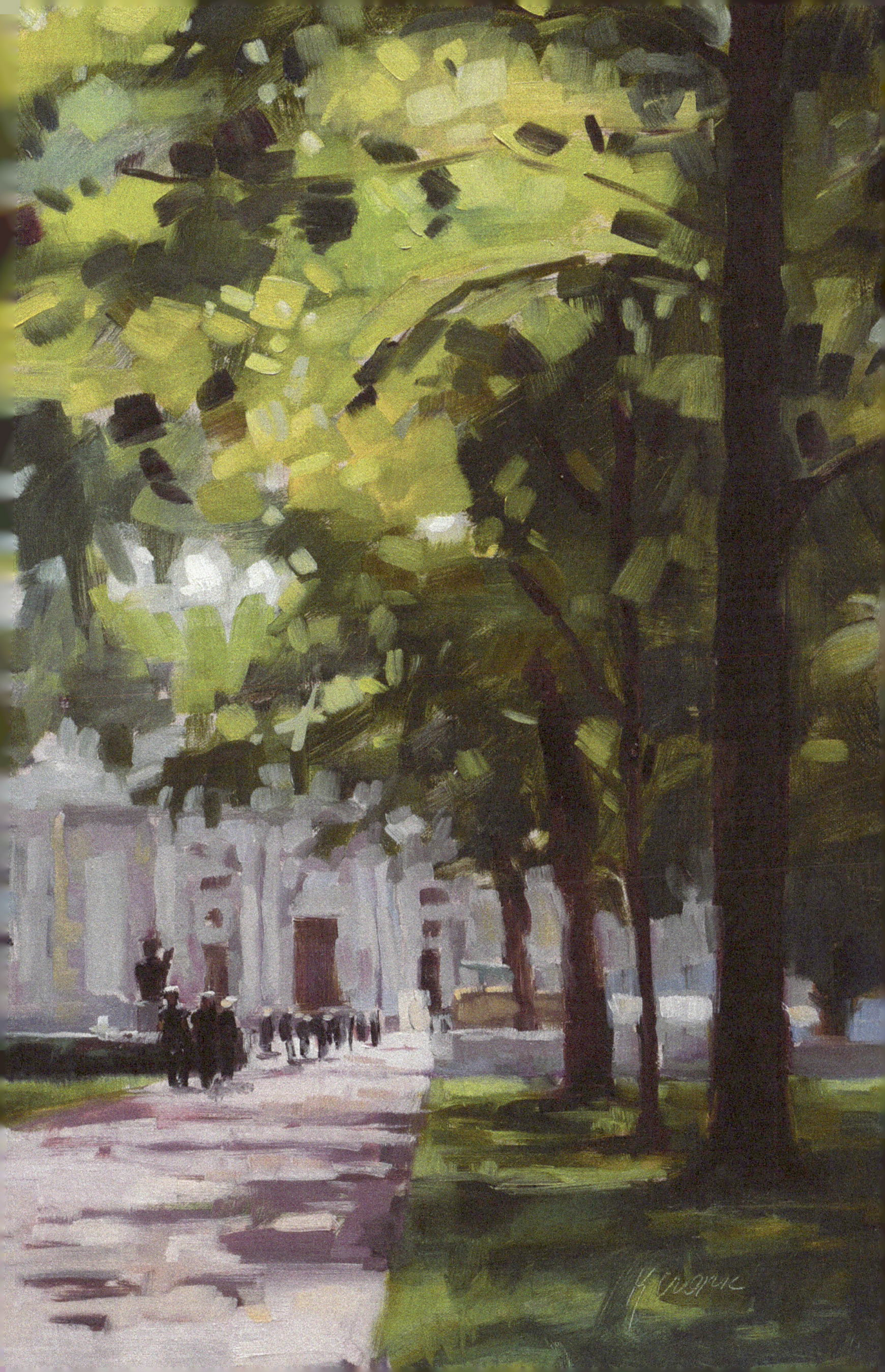

Did you take any electives that you really enjoyed?

Describe what you remember about your final weeks at USNA.

Did you have any memorable Spring Break moments? What were they? Where did you go?

How did you come to service select what you did? Did your preferences change throughout your time? If so, why?

Is there anything you would do different during your time at USNA?

What are you glad you took advantage of at USNA?

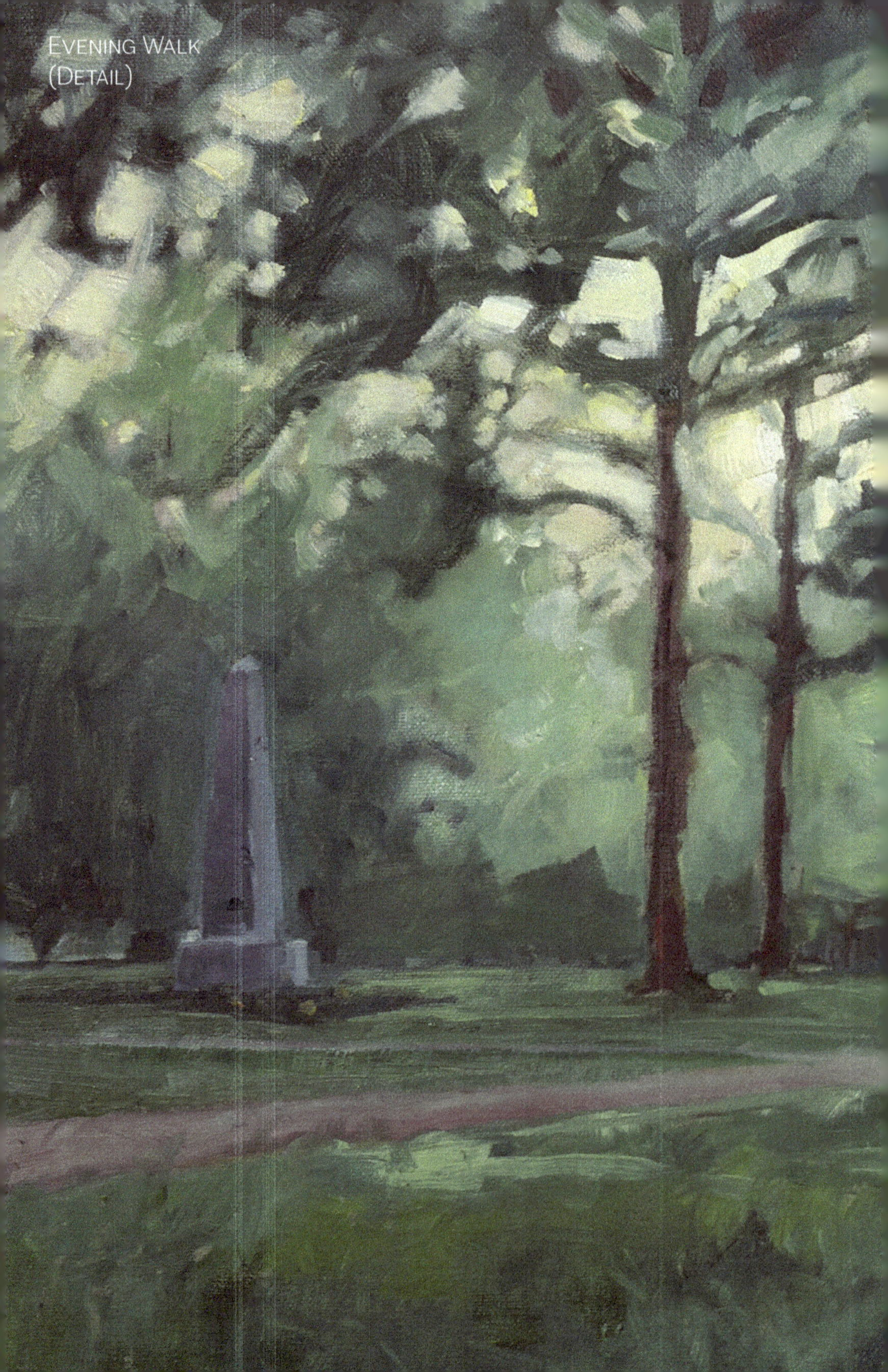
Evening Walk
(Detail)

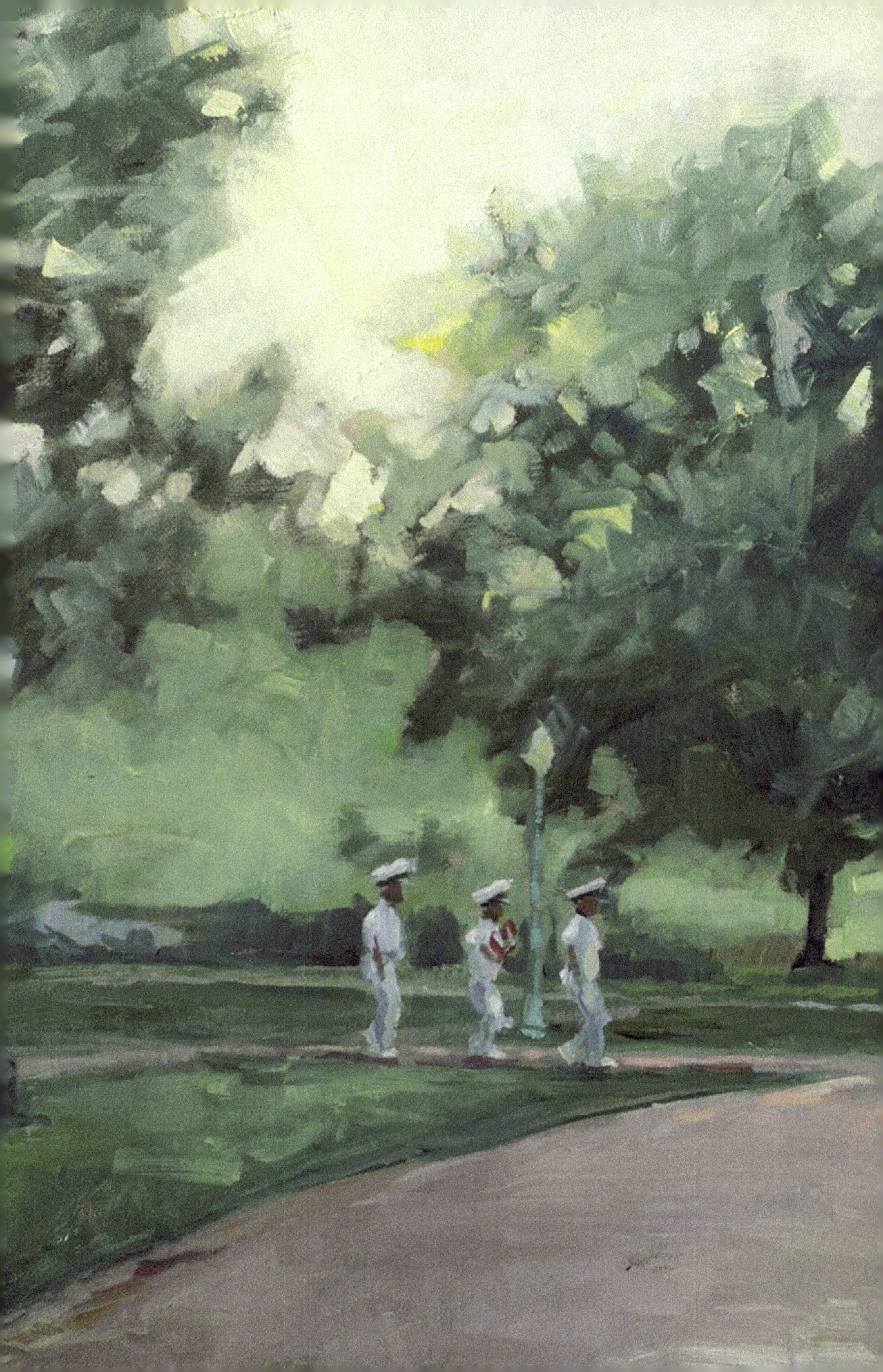

What customs or rituals made you feel proud at USNA?

Describe a moment when you grew personally.

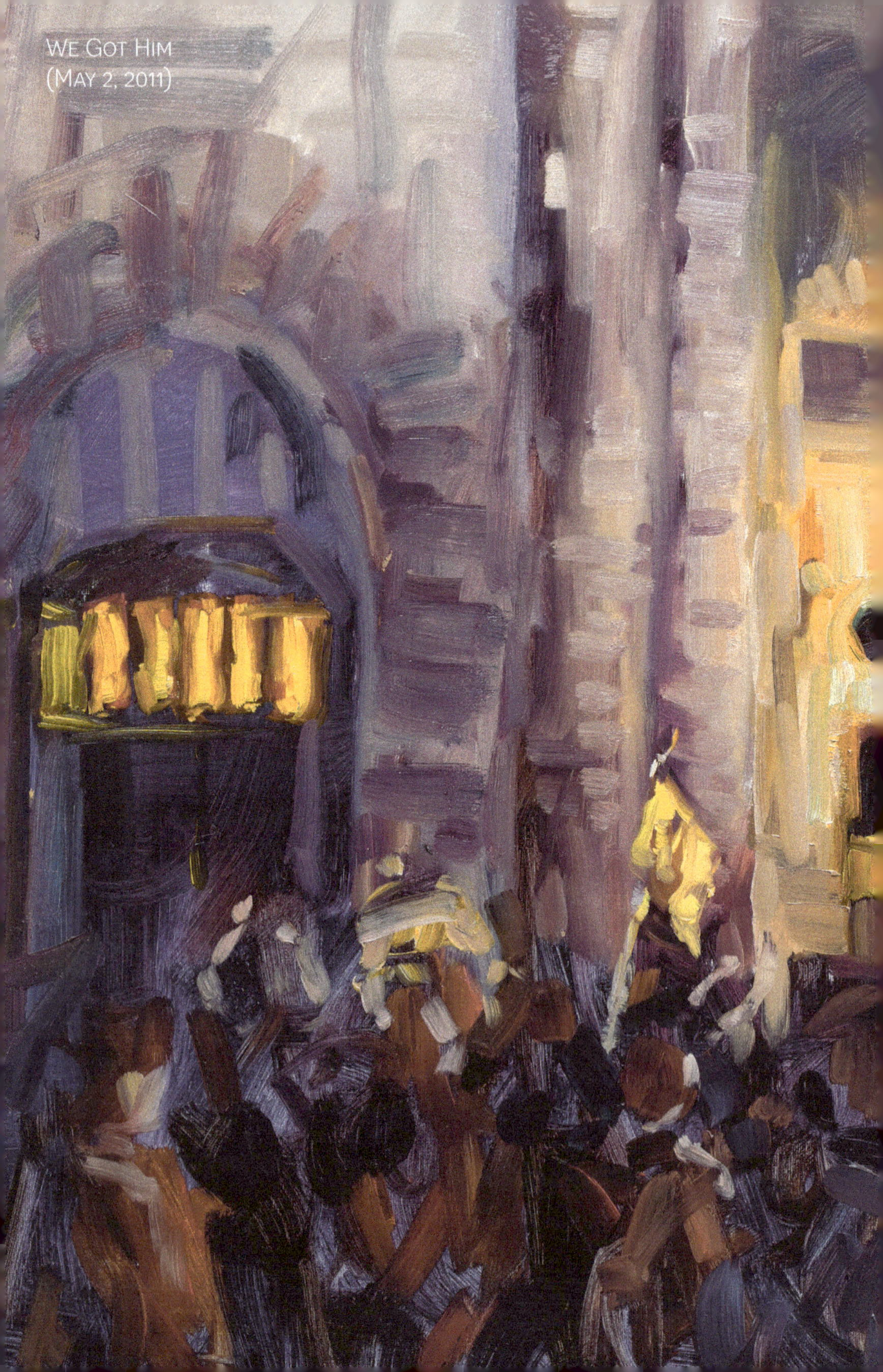
WE GOT HIM
(MAY 2, 2011)

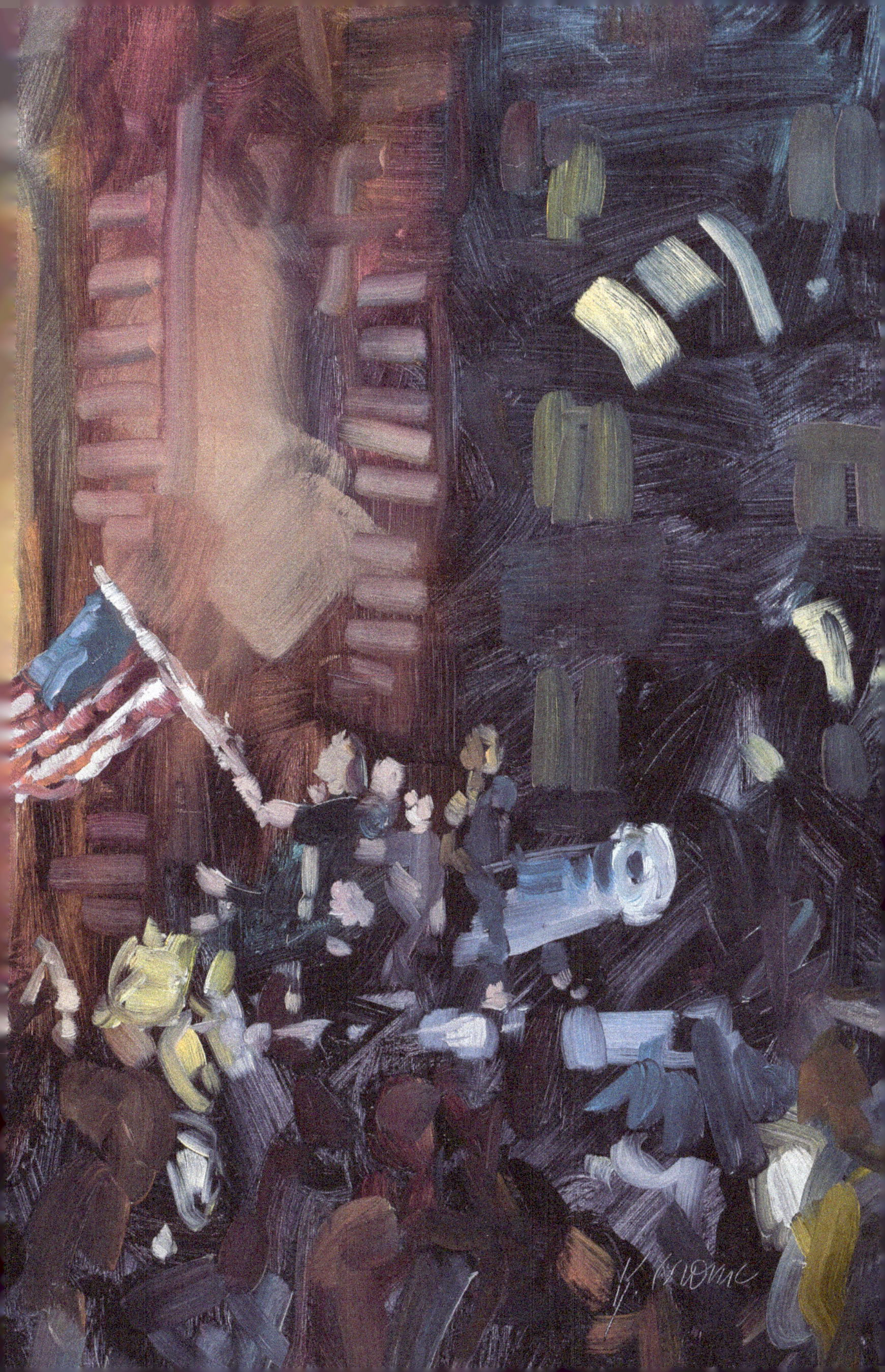

K. Krome

Did anything significant happen in the world during your time at USNA? What do you remember about it?

Do you have any "once in a lifetime" opportunities that you cherish from your time at USNA?

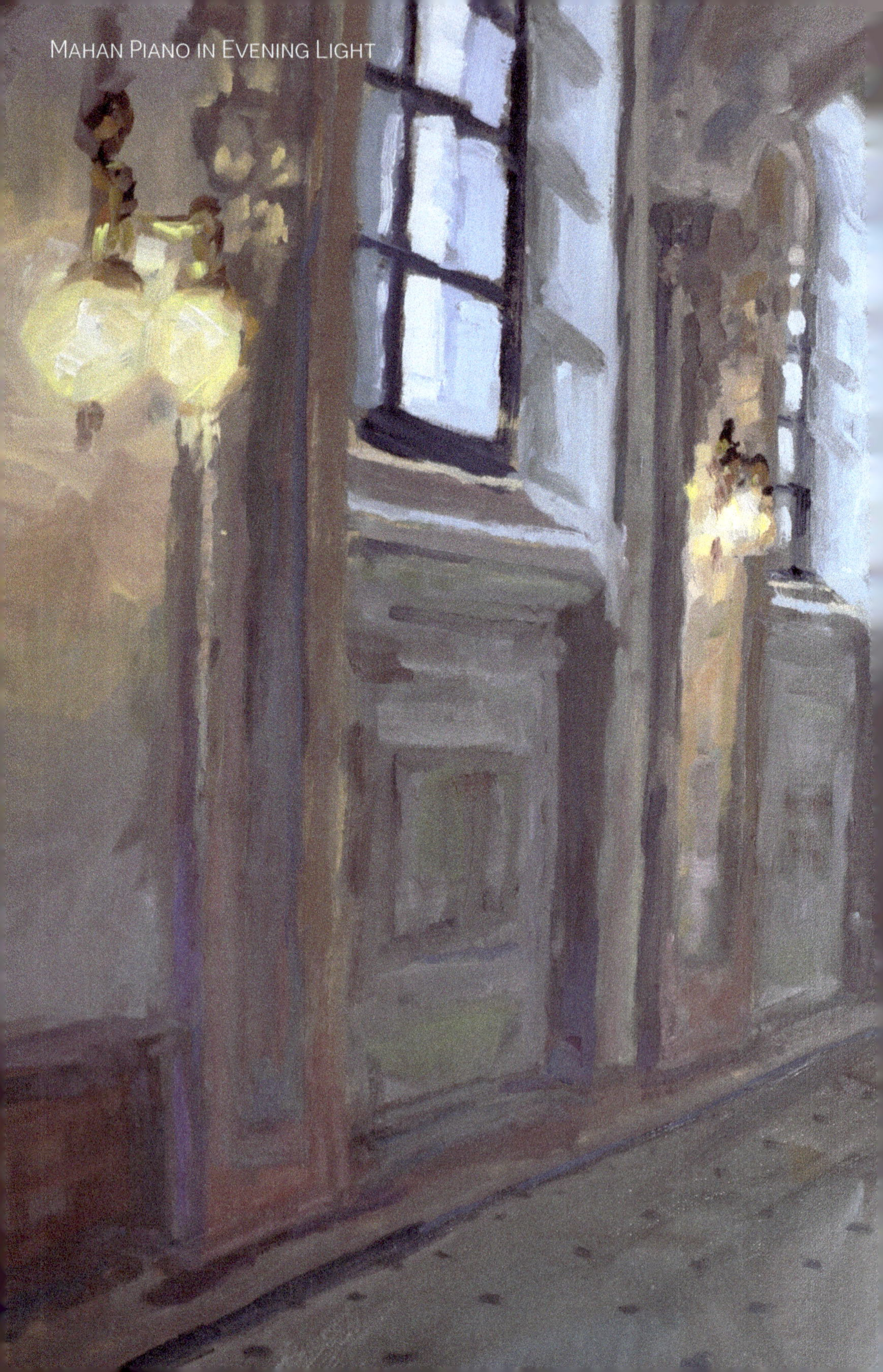
Mahan Piano in Evening Light

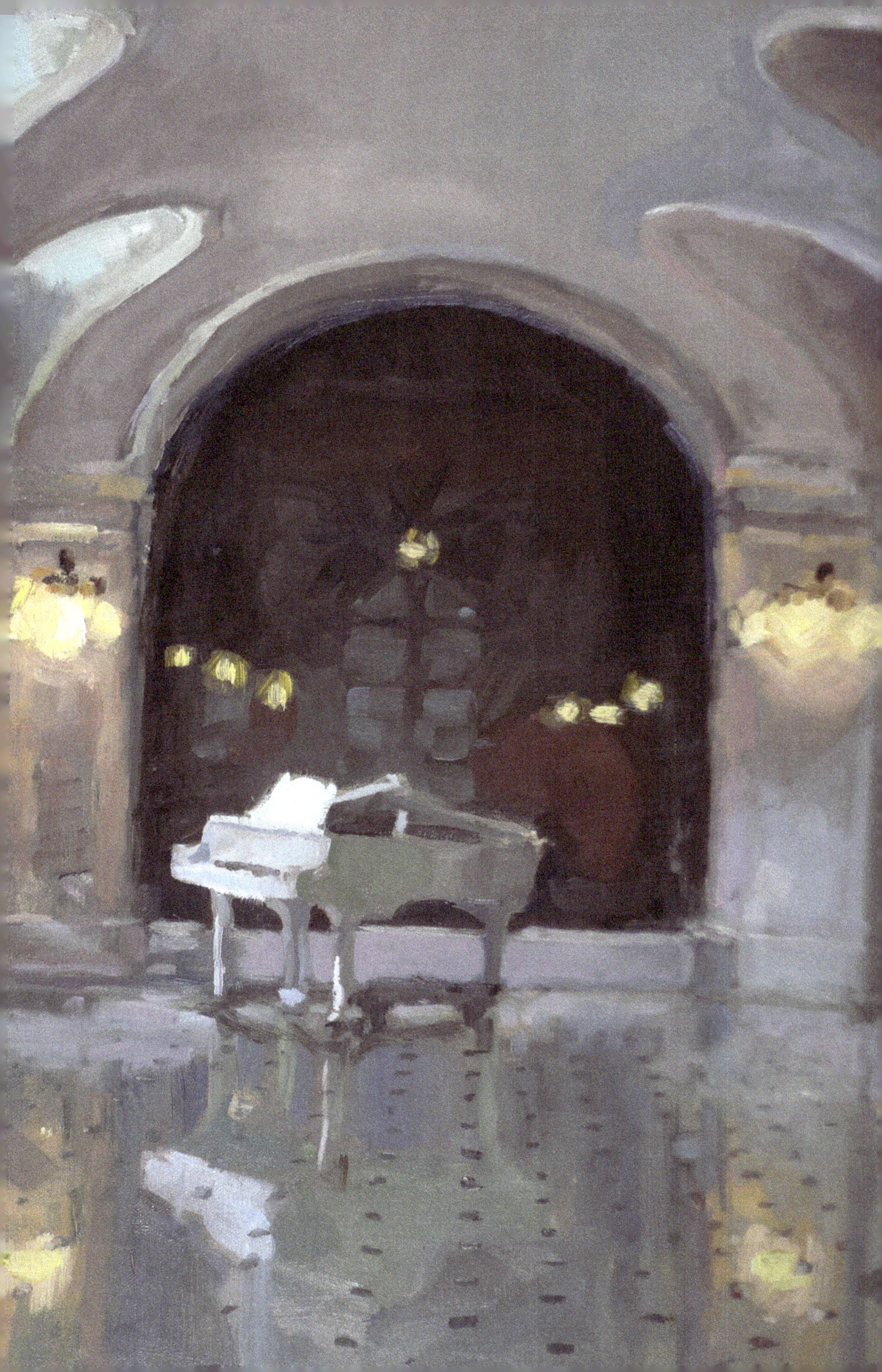

Was your family or friends able to visit you at USNA? What were their favorite parts of the Yard?

What were their favorite parts of Annapolis and the surrounding
area?

How did you feel about USNA when you graduated? Did that change over time? How do you feel about it now?

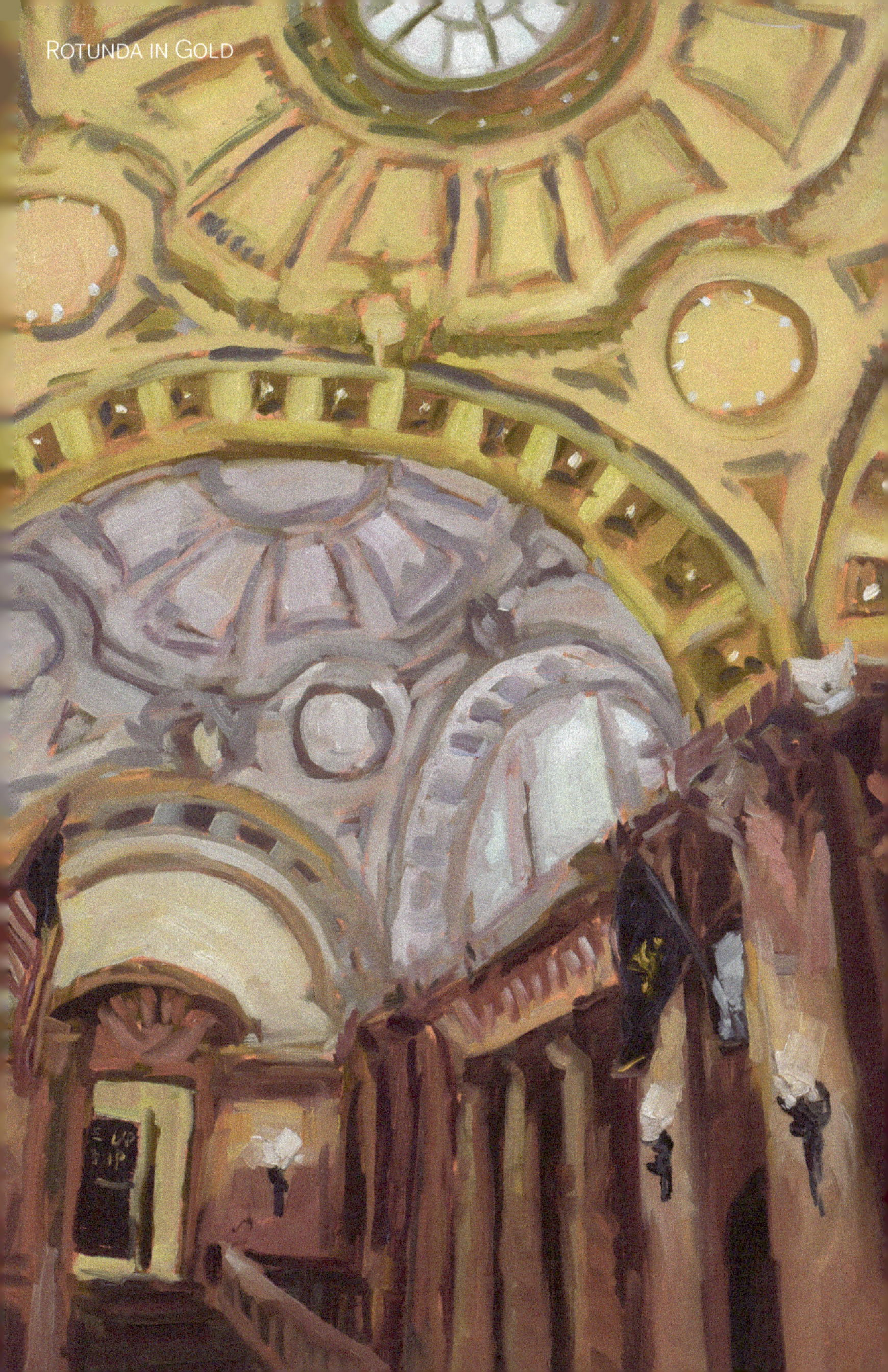
Rotunda in Gold

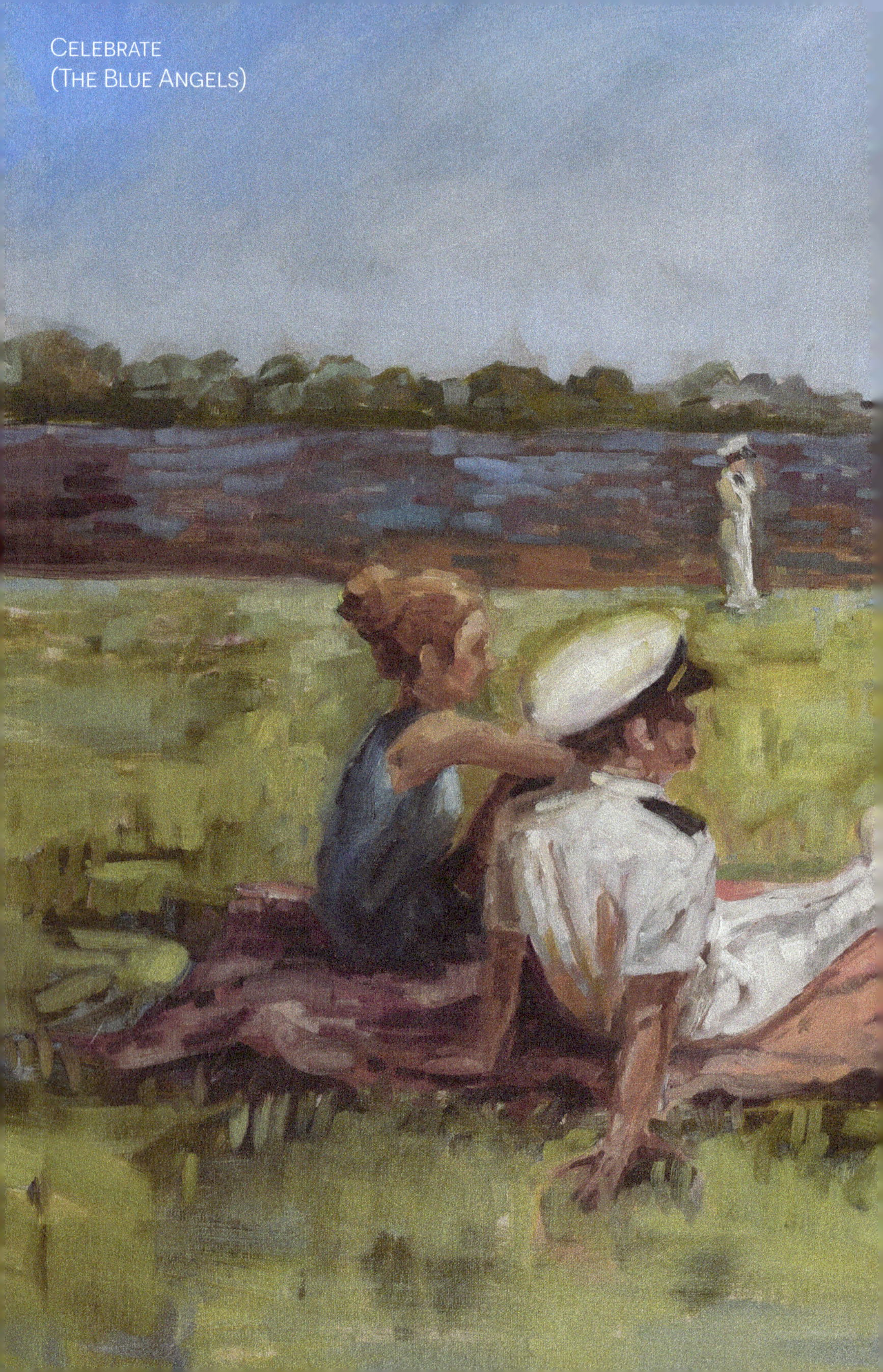
Celebrate
(The Blue Angels)

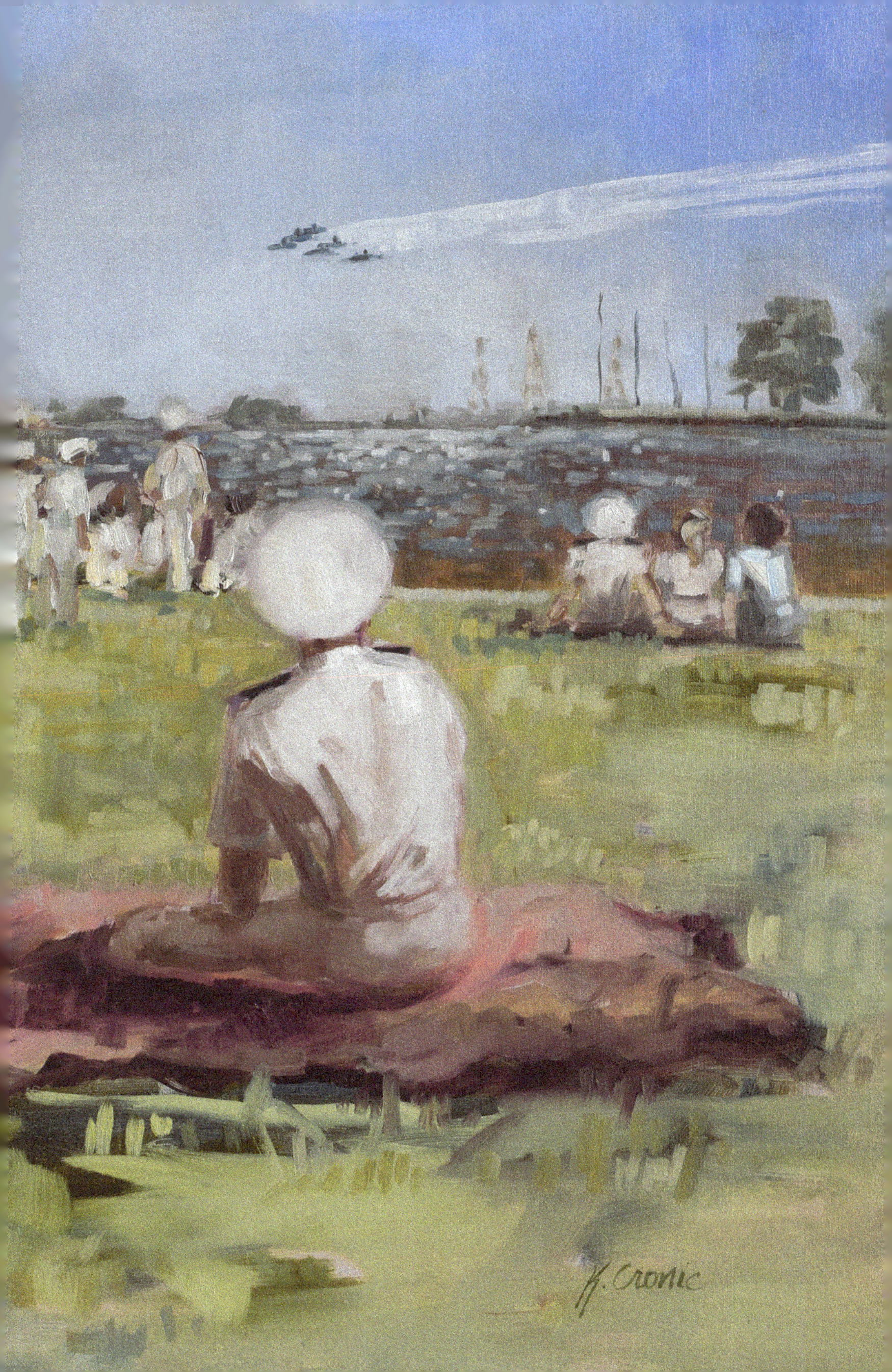

K. Cronic

Share your favorite moments of commissioning week.

Who did you celebrate with? Where did you stay?

Who was your Company Officer and Senior Enlisted Leader
when you graduated?

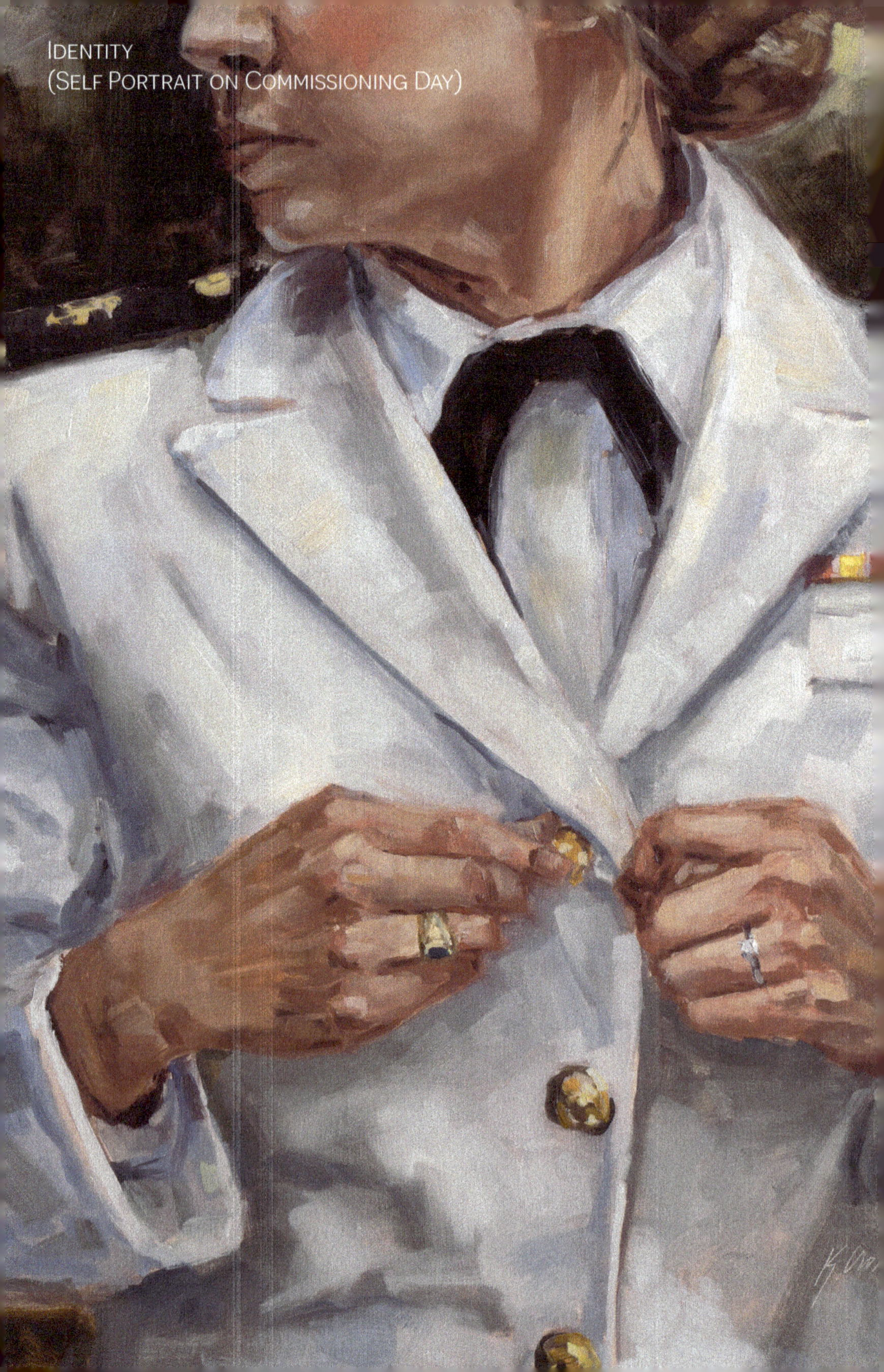

Identity
(Self Portrait on Commissioning Day)

What kind of person did you look like as a plebe? How did that change/not change when you graduated?

What is one challenge you overcame during your time at USNA?

What were you most sad about leaving after USNA?

Have you been back to the Yard? What has changed, and what exactly as you remember it?

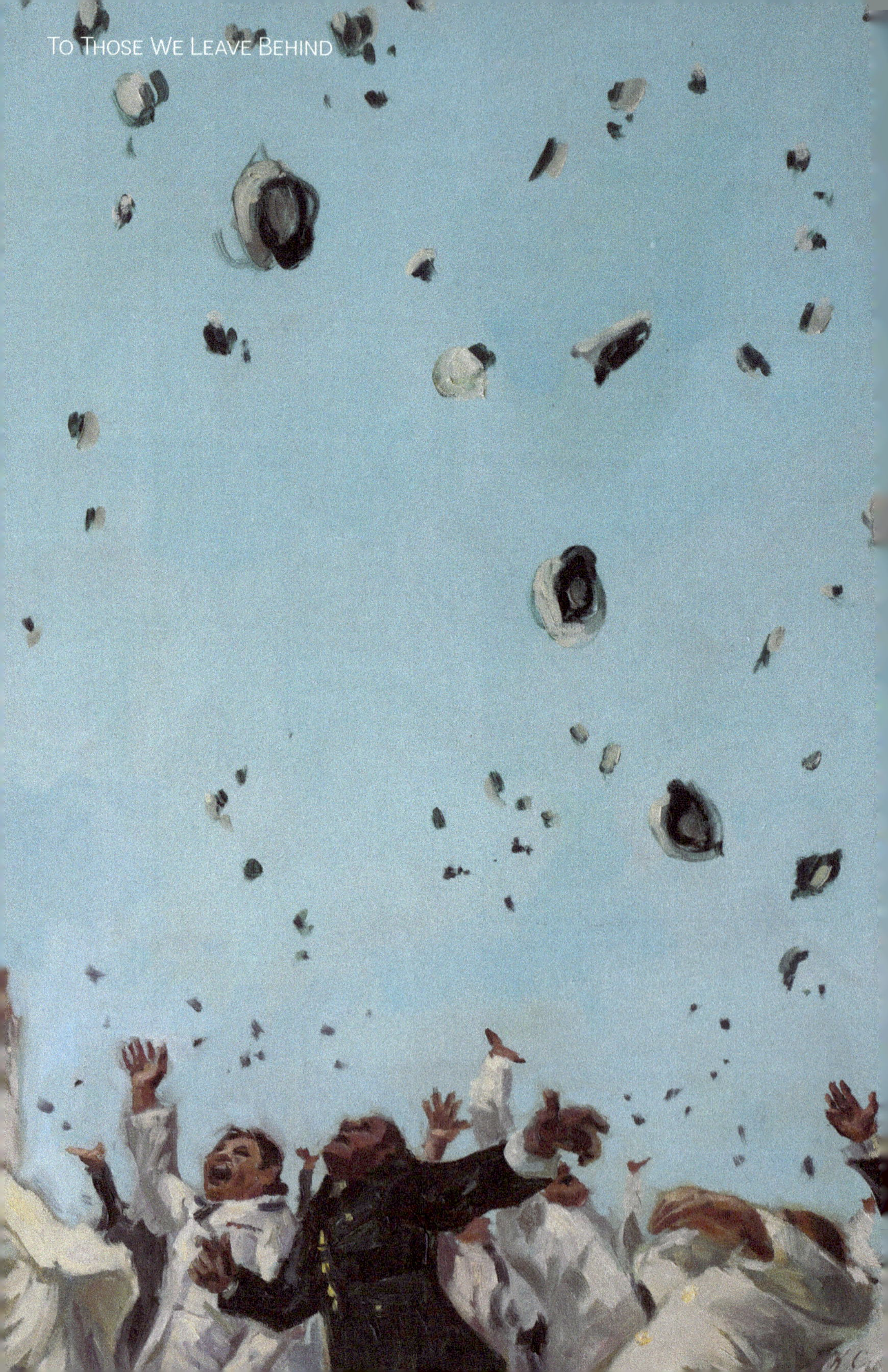

Write down everything you remember about your commission-ing and graduation ceremony. Who spoke? Who were you next to?

Who was your first salute? As an underclassman, were you the first salute for anyone? If so, who?

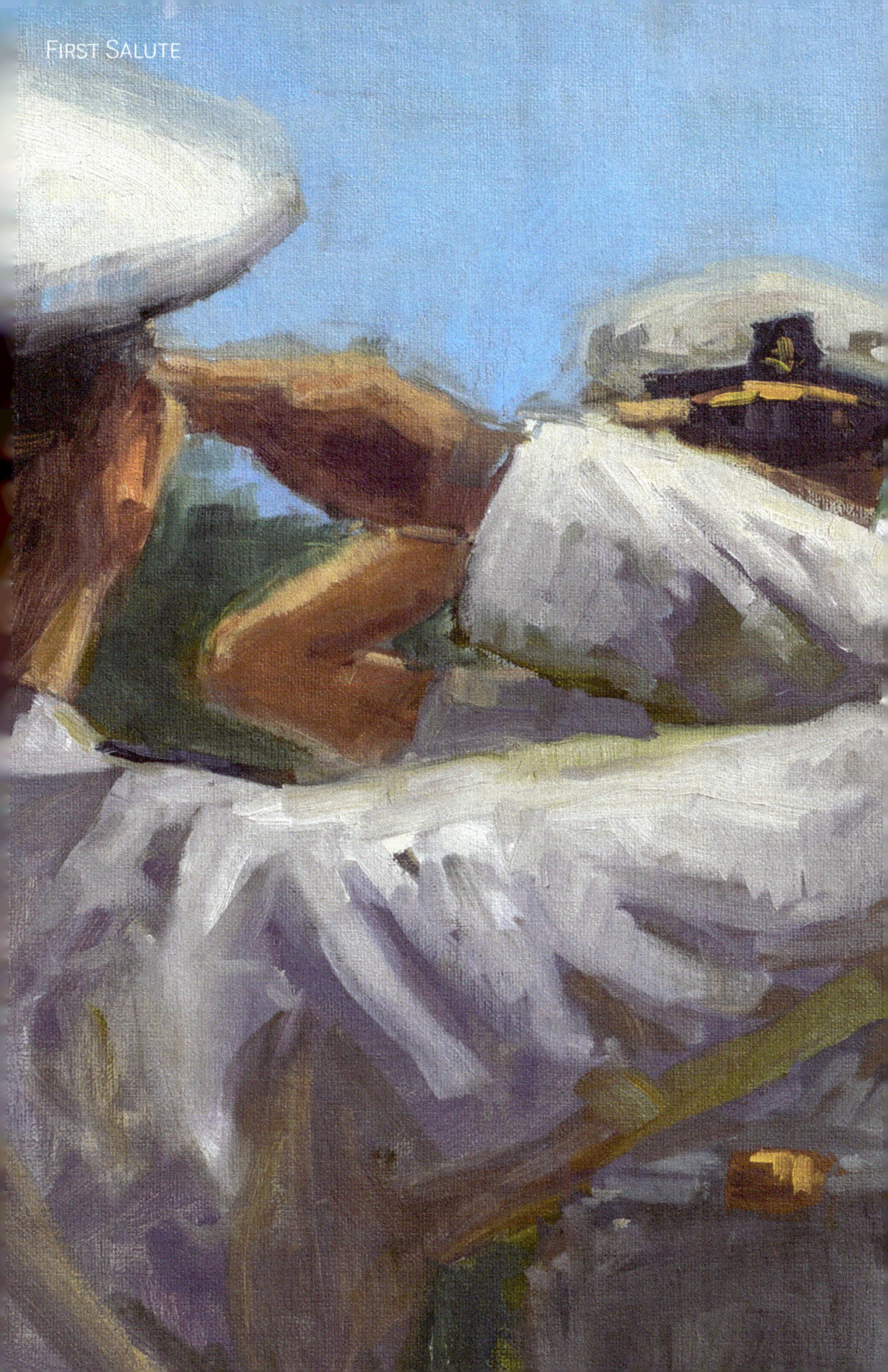
FIRST SALUTE

How did you celebrate your own graduation?

Did you know of or attend any weddings after graduation?

Share one thing you did not expect about commissioning week.

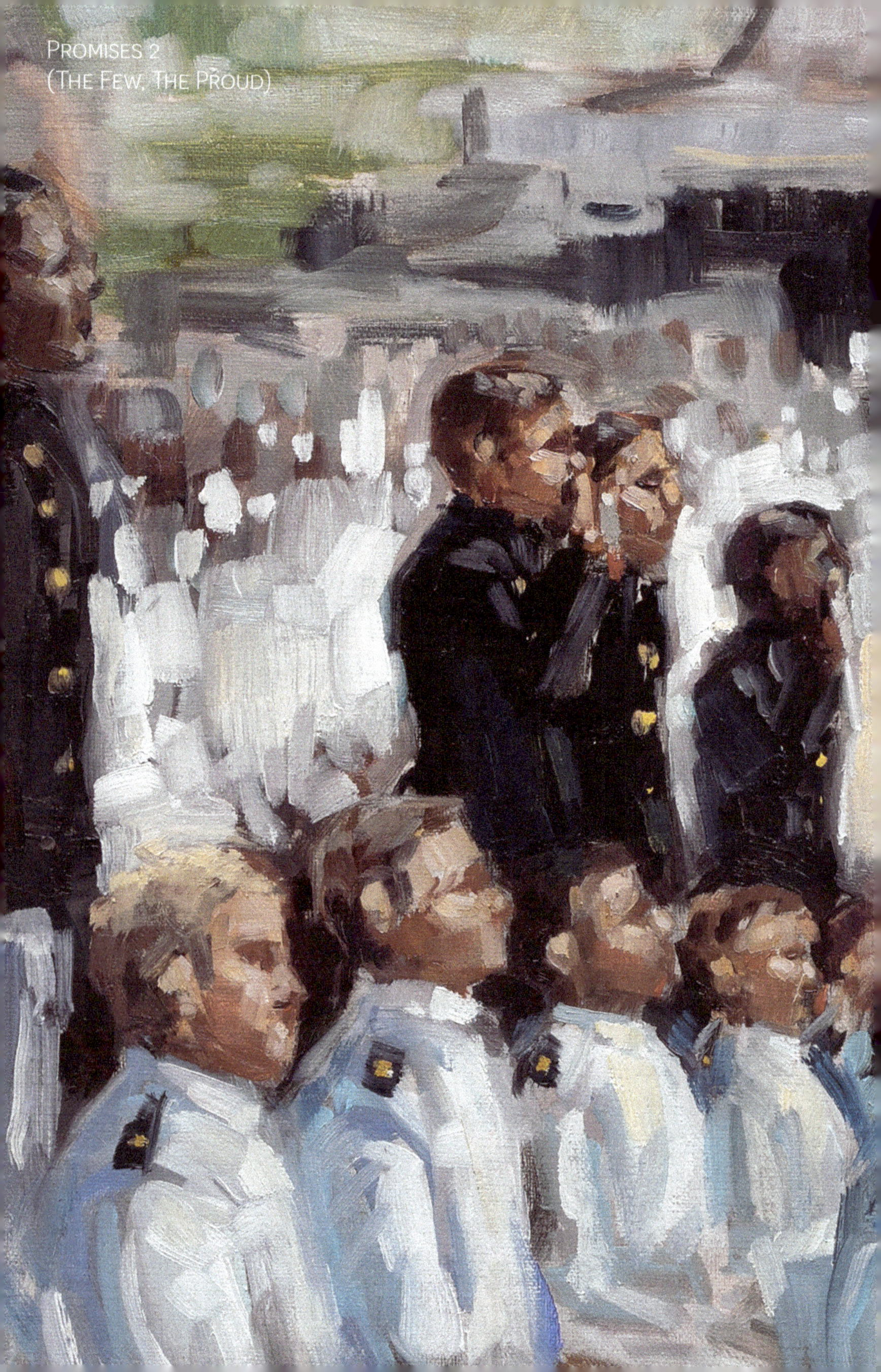

PROMISES 2
(THE FEW, THE PROUD)

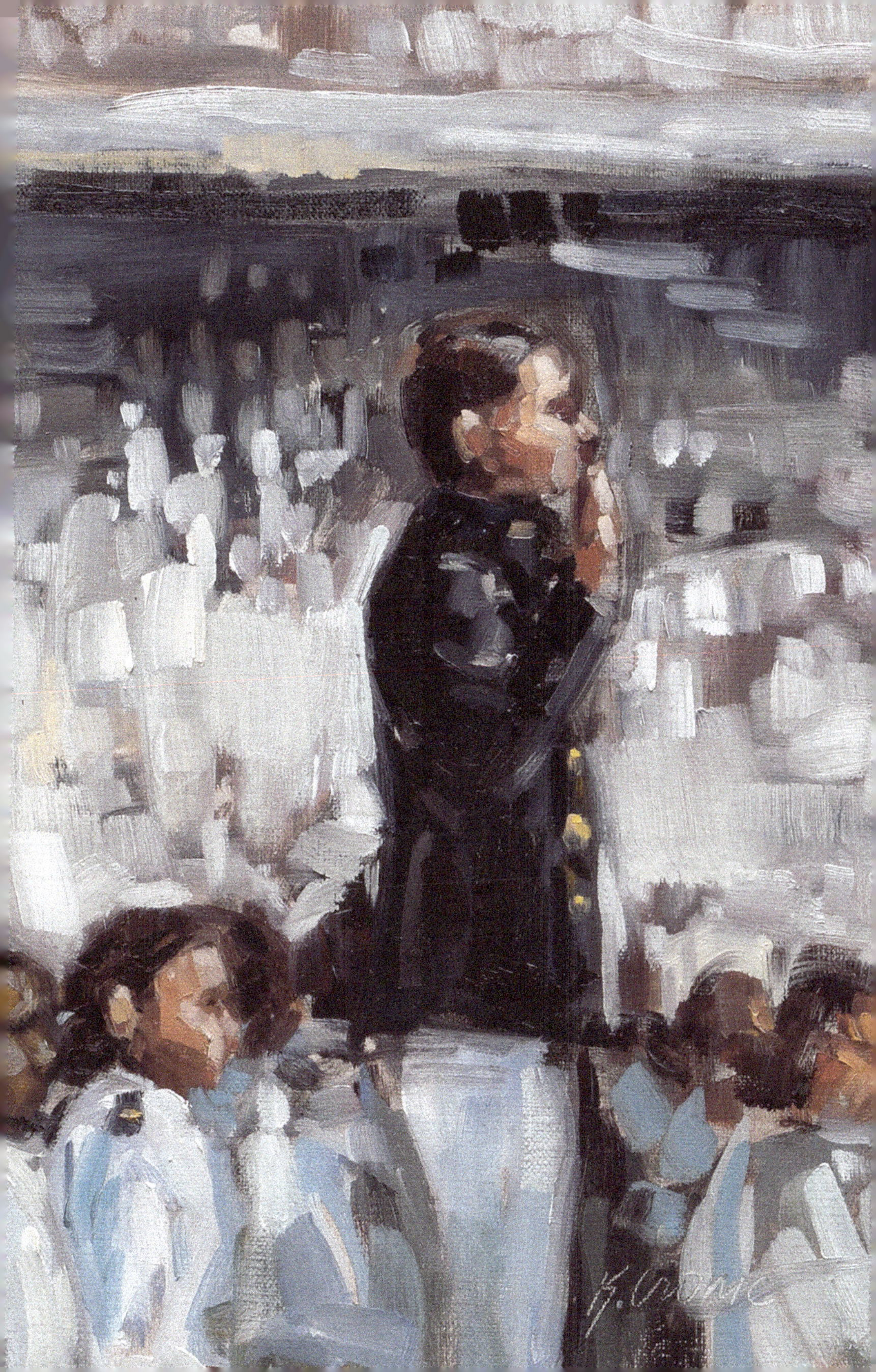

What did your first summer after USNA look like? Where did you live, who were you with, what were you doing?

How did USNA prepare you for your first tour as a junior officer?

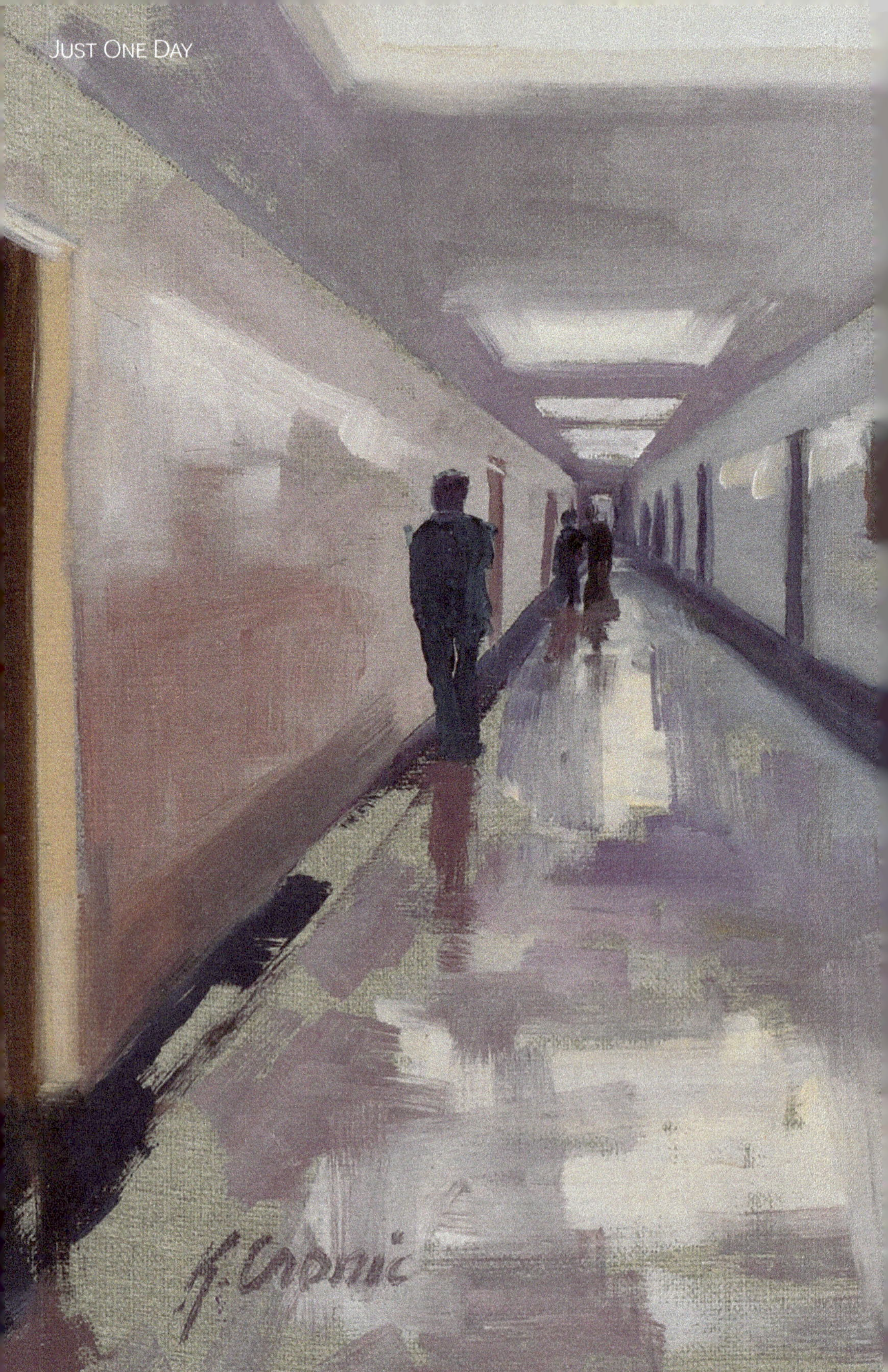
Just One Day

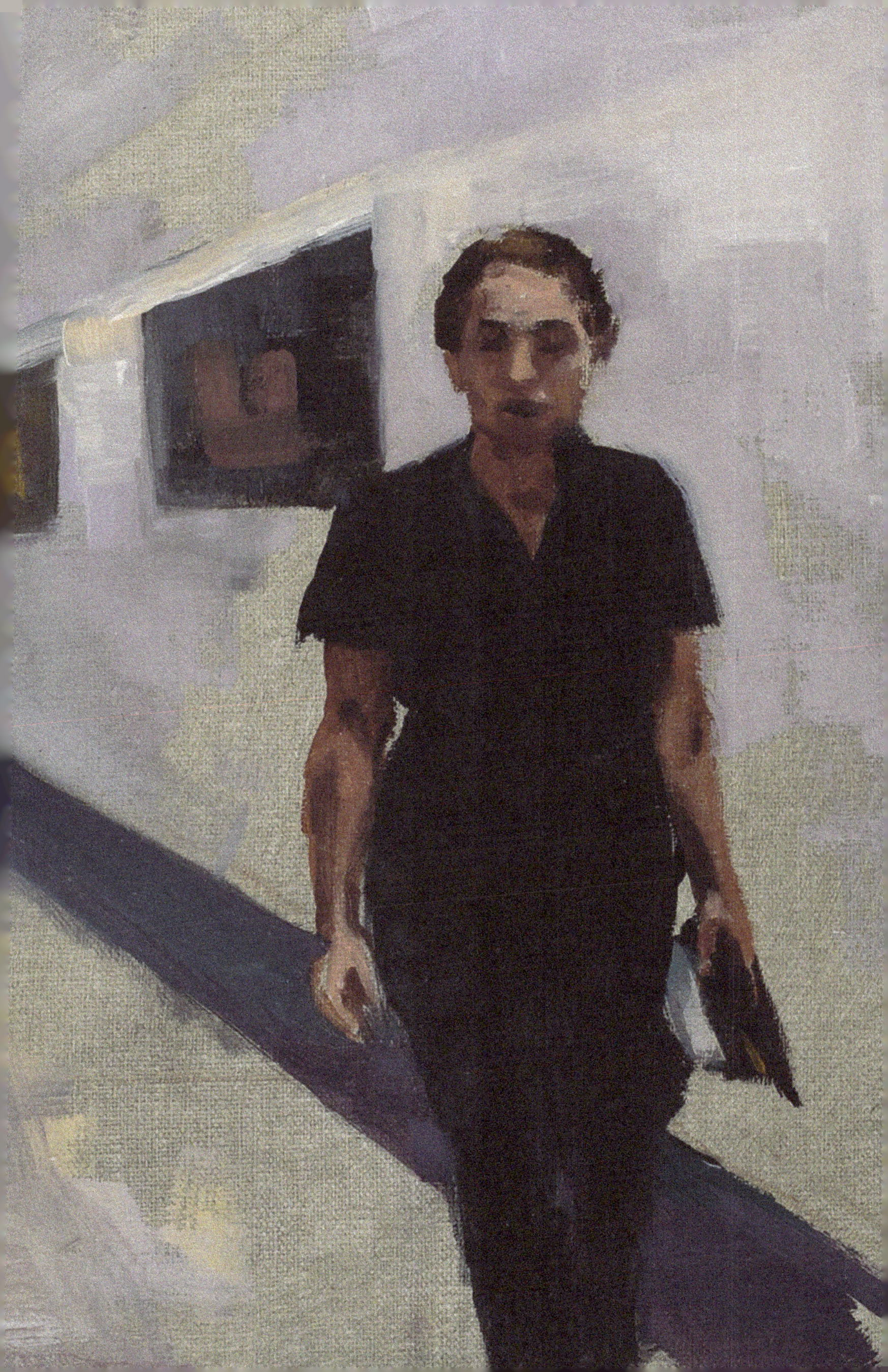

If you could go back for just one day, what would you do? Who would you see?

Under each image, write the
names of friends who service
selected each.

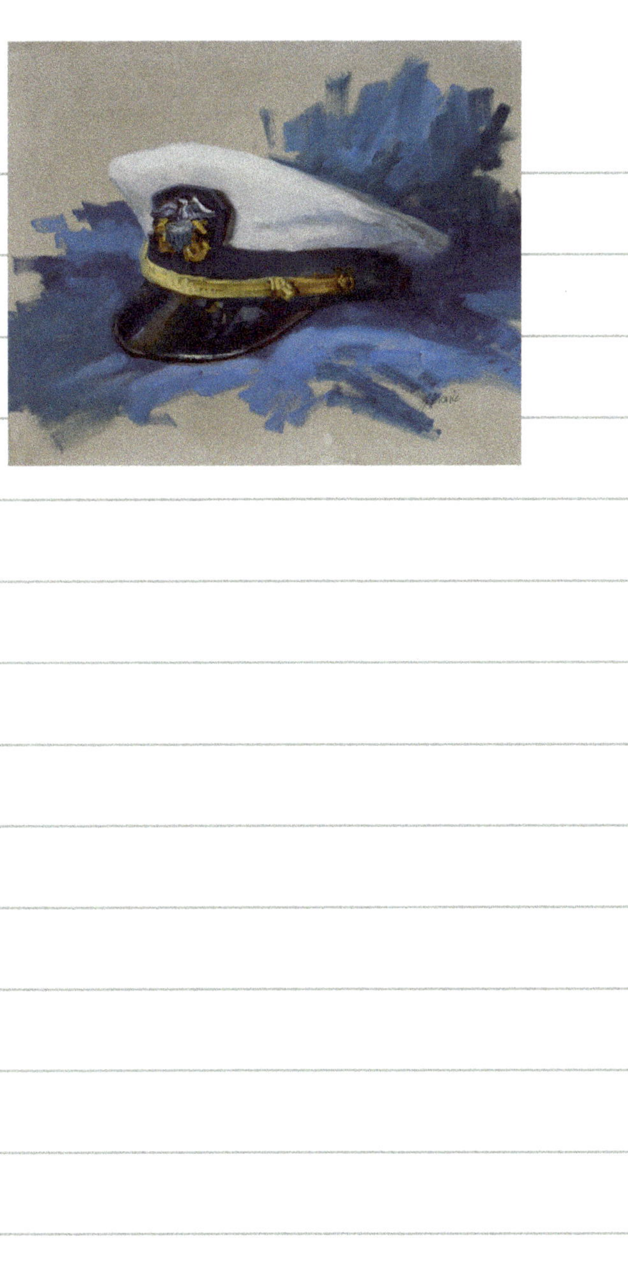

What are they up to now? Note the date and use the following
pages to return to every so often.

Name / Date / Notes

Name / Date / Notes

Name / Date / Notes

Name / Date / Notes

Name / Date / Notes

Name / Date / Notes

Name / Date / Notes

Notes and Memories

Notes and Memories

Notes and Memories

Notes and Memories

Notes and Memories